To Dwight Henry Jr

In appreciation of your
"love of all Nature"

Much Love
Dad & Mother
Christmas 1978

The Fascinating World of the Sea

The Fascinating World of the Sea

Circling the Globe for the Wild Seashell

BY ELVA D. SHEETS

Photographs by Robert Lipe, et al.

Foreword by R. Tucker Abbott

CROWN PUBLISHERS, INC.
New York

This book is dedicated to my children and grandchildren.

It is to thank Cosette, Sumner, and Alice for making one trip possible when the "gasoline kitty" fell two thousand miles short.

It is to thank again my granddaughter, Brenda, for the personal sacrifice involved, giving away six whole weeks of allowance money, "so you will be sure and have enough gasoline money to get home, Grandma."

And it is to assure my granddaughter, Nancy, of the joy she brought with her enthusiasm during her very young years as she helped (?) untie the packages that Grandma brought home.

Inquiries should be addressed to Crown Publishers, Inc., 419 Park Avenue South, New York, N.Y. 10016.
Library of Congress Catalog Card Number: 73–82949
Printed in the United States of America
Published simultaneously in Canada by General Publishing Company Limited

Photographs by Robert Lipe, unless otherwise credited.

Designed by George Hornby

Introduction and Acknowledgments

At age fifty, a quarter of a century ago, it was forcibly called to my attention that people living in the center of the country, a thousand miles from the ocean, were curious and eager to learn about its mysterious behavior and its even more mysterious animal life. So great was the force of that observation that it inspired a vision that has dominated my entire life this past quarter of a century.

With a strong desire to share my own personal fascination, and with an equally strong desire to help others understand the phenomenal natural forces at work in the vast ocean, and to help them appreciate the colorful beauty of its carefully concealed shells and other animal life, I undertook, single-handed, the task of moving the mountain to Mohammed.

Working toward my vision of serving millions of landlocked people, I traveled 185,000 miles, circling the globe in search of colorful shells and other miscellaneous ocean life, all hand chosen for their special appeal to the nature-loving general public. Hoping that my efforts might also benefit shell collectors and scientists, untold hours were spent identifying and cataloging the ocean treasures I found—some twelve thousand different items. But where the interests of the specialist clash with those of the general public, I devote my efforts to the needs, likes, and dislikes of the public, to the thousands who have never before been served with an exhibit of a museum where their needs were given precedence over the needs of science.

To make it possible for everyone to share the collection's unusual colorful beauty and variety, I am making a gift of it, under the name Nature Colorama, to Fort Wayne, Indiana. (A fund-raising drive is now in progress to erect the handsome building designed by architect Orus O. Eash, who is donating his services.)

Assembling the collection, however, was not the entire reason behind the long quarter-century search. It was also my intent to learn and share the compelling story of how the ocean and its creatures work together in a profoundly subtle pattern of mutual dependence, an interaction without which man could not survive. This story—filled with fascinating and unusual nature facts—I am sharing with others through this book. It is written for all adventure and nature lovers, be they eighteen or eighty years old. No previous knowledge of the subject matter is needed for full understanding and enjoyment. Shell collectors, however, will be especially interested by a general discussion of forty-six shell families; the book gives a type of information not found in the usual shell books, which are all written principally to help with identifications.

God grants each of us but one lifetime to live. Had I delayed each collecting trip long enough to save hotel and restaurant money that I might travel in comfort, many lifetimes would have been needed to reach my goal. So, in the hope of completing the project in the one lifetime allowed me, I traveled with only transportation and grocery money. My unorthodox method of travel meant extremes of both rewards and distresses. This book offers the fascinating secrets I was able to learn about the ocean, its animals, and its shells, and a forthcoming volume relates the highlights of my travel and collecting adventures.

My heart is filled to overflowing with gratitude for help received from others. The limited mode of travel intensified my need for, and gratitude to, the many wonderful people who contributed so generously to the step-by-step pursuit of my goal.

First, I want to express my deepest appreciation to the following able scientists who donated many hours of their valuable time checking my manuscript for technical errors: Dr. Harald A. Rehder, Senior Zoologist, Smithsonian Institution, Washington, D.C.; Dr. Gilbert L. Voss, Chairman, Biology Division, Institute of Marine Science, University of Miami, Miami, Florida; Dr. Frederick M. Bayer, Institute of Marine Science, University of Miami, Miami, Florida; Dr. William J. Clench (retired), Director, Museum of Comparative Zoology, Harvard University, Cambridge, Massachusetts; Dr. Fred D. Morgan, Professor of Zoology, Huntington College, Huntington, Indiana; and Mina M. Slinn, Conchology Consultant, Museum of Natural History, Tampa, Florida. I want to especially thank Mina Slinn for her years of loving and generous hospitality, and for her untiring help with difficult identifications. And I want to assure all these generous people that I personally bear full responsibility for any and all errors that may still remain.

Special thanks go also to the following writers and authors for their personal encouragement, suggestions, and advice through the many trials and tribulations involved in making this information available in book form for all to enjoy: Mrs. Harry Inge Johnstone, Mobile, Alabama, Walter Eyerdam, Seattle, Washington; Ormond McGill, Palo Alto, California; Gordon Melvin, West Newton, Massachusetts; and Dr. A. Myra Keen, Stanford University, Palo Alto, California.

Thanks also to the following people who were generous enough to allow shells in their collections to be photographed for use in this book: William Bledsoe, Simon DeMarco, Fred Glancy, Florence Kuczynski, Arch Mellor, Esther Reader, and Mina Slinn.

I thank Dr. R. Tucker Abbott, Delaware Museum of Natural History, for reading galley proof and offering numerous helpful suggestions.

Gratitude and appreciation beyond my ability to express go to family and friends who supported my effort with such good cheer, although I inflicted upon them many weeks and months of worry as I traveled my way through stresses and dangers that to them seemed unendurable. I especially thank them for making it possible for me (for fifteen consecutive years) to transfer all the clothes budget money to the transportation kitty by giving me their mendable, alterable clothes to use.

An extra special thank-you goes to my daughter, Cosette, for making the publication of this book possible. With faith that my material could, and should, be published, she offered to type my manuscript purely on a gambling basis—future payment if success followed, and no payment in case of failure. And a thank-you to my editor, Stephanie Winston, who has worked with me closely in putting this book in final shape.

My hearty thanks to the many policemen, coast-guard personnel, filling-station attendants, and private citizens of widely scattered areas who gave such solicitous care to a lone grandmother who was facing the big bad wolves of the world as she pursued her dream. And a special thank-you to foreign missionaries who offered a haven of refuge in strange lands, and to the dark-skinned doctor and dentist who served so capably without aid of a common language.

To the many divers, dredgers, shrimpers, trawlers, deep-sea fishermen, beachcombers, shell collectors, shell dealers, and shell clubs go my undying gratitude for their invaluable help. Their personal observations added tremendously to the interest value of this book, and without their help it would have been quite impossible to build the collection as it now is. Another special thank-you goes to "Sissie" Kern of Kahului, Maui, a member of the Hawaiian Divers Club. She arranged a shell shower party in her home for me. Those wonderful strangers of Hawaii sent me back to the mainland with over a thousand dollars' worth of their beautiful deepwater shells for my exhibit. For their substantial contribution of information, appreciation also goes to the many libraries, museums, aquariums, professors, pharmacists, and marine biologists.

I would especially like to acknowledge the help of these museums:

Inside the United States:

Smithsonian Institution, Washington, D.C.
Field Museum, Chicago, Illinois
American Museum of Natural History, New York, New York
Museum of Science and Natural History, Tampa, Florida
Harvard University Museum, Cambridge, Massachusetts
Stanford University, Palo Alto, California
Rollins College Beal-Maltbie Museum, Winter Park, Florida
Natural History Museum, San Diego, California
Natural History Museum, Los Angeles, California
Natural History Museum, Olympia, Washington
Museums of Hawaii

Much help was obtained from smaller museums including those located at Fort Myers, Florida; Abbeville, Louisiana; Corpus Christi, Texas; Beaufort, North Carolina; Beaumont, Texas; Houston, Texas; Arden, North Carolina; and many more.

Outside the United States:

The large natural history museums at Sydney, Australia; Wellington, New Zealand; and Auckland, New Zealand.

My sincere thanks also to the hundreds who wrote helping me plan collecting trips in their respective countries, many of whom were so wonderfully helpful after I arrived—often working on through the night helping with identifications, and packing to meet a departure date. I want to especially thank the troops of delightful children who gave such enthusiastic help on the beaches, especially those on the Fiji Islands, Mexico, Africa, and the Eskimo children along the Bering Sea.

And last but not least, I want to thank the Almighty hand that seemed always to be there at a time of greatest need.

ELVA D. SHEETS

EDITOR'S NOTE ON THE AUTHOR'S SHELL "PICTURES"

The three shell design photographs—Plate 1, Plate 13, and Plate 25—show artistry in a new medium; the medium, not of paint or stone or wood, but of seashells. The arrangements, not to be confused with shell craft, were created by Mrs. Sheets from her original designs. She carefully explains that the Great Creator, and not she, is the artist who painted the beautiful colorful objects, and that she merely collected and arranged His treasures for people to better appreciate their beauty.

The circular picture is seven feet in diameter, and the two rectangular ones are five feet tall. They are comprised of shells from shelling areas of many different countries of the world. Drawing on her entire collection of land and tree shells, as well as ocean shells, she spent many months working out her designs by the trial and error method, with no pattern to use as a guide.

It was a matter of trying many combinations until she finally found shells of the right texture, color, and size that could be grouped harmoniously together for color emphasis; often an arrangement idea had to be abandoned for want of a few extra shells of a certain species. In most mediums, running short of a certain material only means a trip down to the supply store to buy a little more. But if one needs even one more shell to complete an idea, the whole idea must be abandoned; the supply store doesn't carry the materials one needs. Nor could the author travel 10,000 miles across the world to pick up just one more of that particular species, in the one place in the world that that species could be found. It is equally impossible, with most species, to collect shells of equal size. Since all sizes must be used, the constant and tedious problem of size graduation must be solved as each shell is placed. Imagine scraping all the white pectens bordering the black panels of the round picture piled helter-skelter in a box, and then coping with the unlike sizes as you replace them in symmetrical graduated rows. Or imagine the same size-gradation problem on the rectangular picture with the scroll-work, when you are faced with a big box piled helter-skelter with all those hundreds of different sized "coffee" shells.

Those familiar with ocean life will recognize that Mrs. Sheets used not only shells, but also chitons, chiton sections, turbo operculums, and *Spirula spirula*. Spines from a slate pencil sea urchin were used for the brown stars at the outer corners of the white panels on the round picture.

The pictures are presently on temporary display in Fort Wayne, Indiana, where a building project is in progress in which they are to be permanently displayed, along with Mrs. Sheets's entire Nature Colorama collection.

Foreword

This book on marine biology is by a dedicated nature lover, as opposed to a true scientist—as Mrs. Sheets will tell you herself. The book is remarkable because it shows what grit, devotion, and passion toward oceanography and conchology can do to the life of a retired Midwestern lady. Shell study has usually been called the "Queen of the Natural History Sciences," but I think it is time for it to be unmasked as *La Science fatale*.

While a young girl, before World War I, Elva Sheets studied science at what was then called the Bradley Polytechnic Institute (now Bradley University) in Peoria, Illinois. She then proceeded to rear a family, and it was not until the midpoint of her life that she suddenly and accidentally became "shell-shocked." Since then, in 1948, she became obsessed with the pursuit of shells, traveling to such outlandish places as Polynesia, India, and the Great Barrier Reef of Australia.

Then, the indomitable Elva had a revelation to pass on her discoveries of the joys of self-education and the thrill of exploring the mysteries of nature, the sea, and especially the kingdom of the seashell. Undaunted, she has launched into establishing a nature museum in Fort Wayne, Indiana. This book is part of her crusade, which in fervor can compare only with that of Mary Baker Eddy of Christian Science fame and Mrs. Carry Nation of temperance reputation.

One of the charms, and perhaps greatest contributions, of this book is in the simple and straightforward manner in which the mysteries of the sea have been presented. From a basic explanation of tides and waves, to a panoramic coverage of all the well-known sea creatures, Mrs. Sheets has given the new student of biology and oceans an interesting and entertaining account. Woven into the tales of science are many humorous personal anecdotes that enliven her history of oceanic life. Whales, sea urchins, starfish, corals, seaweeds, and crabs are all interestingly covered, but the greatest emphasis, and the one subject that occupies the author in her own personal life, is on marine shells.

This is a book about a person and her exciting discoveries; it does not pretend to be a scientific treatise on malacology or marine biology. And, indeed, the book only goes to prove that "how you say it" and with what enthusiasm and devotion you express your interests may often be more satisfying than the technical precision of actual scientific coverage. Have fun reading it! And, like Robert Louis Stevenson, who said it was better to have an interest in conchology than to be born a millionaire, I hope you will now be condemned to a life of admiring shells and the natural world around us.

R. Tucker Abbott
du Pont Chair of Malacology
Delaware Museum of Natural History

Contents

The Ocean and the "Tree of Life"

1. About the Ocean

Many eons ago our sun, solar system, and oceans were formed, along with everything else, out of the basic matter of the universe—no one knows just how. Each century old theories are discarded and new ones are advanced as scientists become more skillful and their tools more intricate. With their increasing skill, it is conceivably possible that, by the end of another twenty centuries, that most fascinating mystery of all time will be solved. Meanwhile, it is not my intent to probe into those still unsolved mysteries of origin. In my devout love of nature I want only to share the great joy I find in studying a marvelous plan of life, a plan of life that reaches far beyond the ability of mere man to originate, plan, perfect, and even understand.

Since the ocean as it now exists is an undisputed fact, however it got there in the first place, and since it is known to be the source of all existing life, what better place could we start the unraveling of our nature story than by taking a look at it through the eyes of our early ancestors?

Surrounded by sand and soil, with only very primitive means of transportation, our early ancestors had a total misconception of the ocean. Believing it to be a narrow river flowing around the outer edge of the world, they misnamed our planet Earth. It should far more properly have been called Ocean, after the areas of salt water covering 70.8 percent of its surface.

Of all the planets that encircle our sun, the earth is believed to be the only one covered by oceans. In fact, it is believed to be the only one with a surface temperature that falls between the freezing and boiling points of water, so that a liquid ocean is possible. The high capacity of water for storing heat plays a vital role in stabilizing land temperatures to make the earth habitable.

The ability of water to dissolve other substances also plays a vital role in making life possible on earth. Neither land nor ocean life could exist but for that solvent quality of water. All living organisms are but small chemical factories that live, grow, and perpetuate life by means of chemical reactions, reactions that can occur only because the reacting substances are dissolved in the waters of plant and animal tissues. No dry leaf, root, or stem could function in the production of food for the plant, and no dried-out animal could function in any way.

Ocean plants and animals take the moisture that sustains them directly from the ocean, while the moisture in the tissues of land plants and animals comes indirectly from the ocean by the evaporation-rainfall route. The vast ocean surfaces saturate the air above with evaporated moisture, which then condenses over the land into rain. Without this process, all land surfaces would become arid and incapable of sustaining life. Thus, all living things, both on land and in the sea, owe their life to the vast oceans.

Down through the centuries the same water repeatedly makes the same circuit over and over again—evaporating from the ocean, forming clouds, then back to earth in the form of rain. All the rain that falls on land returns by the force of gravity to the ocean, carrying with it a load of salt and other minerals dissolved from the soil. Besides salt, sixty other chemical elements are dissolved in the waters of the ocean. They are washed into the sea by rivers that grind down mountains and wash away plains on the continents. Some elements come from rocks beneath the ocean floor, and others from volcanic ash and windblown dust. Only the water evaporates when it begins its next circuit, since minerals are incapable of evaporation, and the minerals are left in the ocean. Thus, with each circuit the ocean becomes a little richer and the land becomes a little poorer in salt and mineral content. All the ocean's salt would cover the continental landmasses with a layer five hundred feet thick.

Many chemicals of the sea are now used commercially. But due to the hostile environment and high costs, scientific ocean research moves along at a discouragingly slow rate. Seawater is known to stimulate healing, although for some wounds it is harmful. It is now believed that the healing property may come from fungi rather than minerals. One strain of fungus *(Cephalosporium acremonium)* yields two antibiotics, marketed as Keflin and Loridine, which deal effectively with penicillin-resistant staphylococcus bacteria. Seawater has other medicinal properties as well. It provides the bromine used in many sedatives. The metal magnesium, processed from seawater into drug form, is used as a laxative and sold under the trade name Milk of Magnesia.

Thousands of researchers are needed to uncover even a fraction of the sea's medical secrets. So far, barely a scratch has been made on the surface of possibilities.

The ocean floor has huge mountains and valleys that dwarf those on land. Mount Everest would sink down into the Pacific Ocean's Marianas Trench with 6,772 feet to spare. With modern deep-sea research vessels, still more is being learned about the bottom of the sea. In certain areas beds of sediment twelve thousand feet deep have piled up through the downward drift of debris over centuries. Beneath the sediment, there is a heavy rock called basalt, a much harder substance than the granite rock that underlies the continents. In certain areas potato-sized nodules lie on the ocean floor in unbelievable numbers. The nodules, which are composed of valuable minerals, have been lying there for centuries, waiting for ingenious man to discover, salvage, and utilize them. Engineers are now at work building huge vacuum cleaners with three-mile hoses to sweep up the millions of dollars still waiting in those nodules at the bottom of the sea.

The fascination that the sea has always exerted over man's imagination stems in considerable measure from the fact that it pulsates, that it almost seems to breathe like a living thing. Whatever else may change in the world, this rhythm—the pull of the tides, and the roll of the waves—is one constant. So let us take a look now at these tides and waves, and at the shorelines they mold.

Tides

As far back as Roman times, Pliny, in A.D. 77, recognized that the tides had some relationship to the movements of the sun and moon, but the relationship was not understood. The tidal movements were charted, however, and the British Museum has a tide table dated A.D. 1213 that shows the expected dates of the "flood" in relation to the number of days after the new moon.

From then onward, with increasing knowledge of astronomy, more and more relationships were noted between the variations of the tides and the rhythmic movements of the sun and moon. But the underlying cause still remained a mystery.

Not until 1687 did Sir Isaac Newton conceive the first workable plan for tide study, based on his universal law of gravitation: There is a force, called gravity, which tends to draw all bodies in the earth's sphere toward the center of the earth. Since then much progress in understanding has been made, but there is still much to learn.

It is now known that the rise and fall of the tides is due to the gravitational pull of the sun and moon.

For a simple mental picture of worldwide tide action, just think of an enormously long bulge, or swell, two or three feet high that extends directly underneath the moon for the full length of the ocean. This bulge, technically called a tidal wave,* is slowly pulled by the orbiting moon across the ocean and on around the world. Then think of a second, somewhat smaller, long bulge on the exact *opposite* side of the earth from the larger tidal wave. This swell, which is called the centrifugal bulge, is formed by the outward-throwing force of the rotating earth at the point where the moon's pull is weakest.

While the moon slowly pulls the tidal bulge on its trip across the ocean floor and around the world, the somewhat

* Many people mistakenly believe the term *tidal wave* applies to the huge destructive waves resulting from undersea earthquakes and volcanic eruptions. Such waves are actually called tsunami, and will be discussed later in this chapter.

smaller centrifugal bulge on the other side of the world moves forward all the while at the same speed to maintain its position exactly opposite the moon's strongest pull.

Any point has two high and two low tides every twenty-four hours. When the first bulge, or tidal wave, passes over a given area, that area has its first high, or flood, tide of the day. Then approximately twelve hours later, when the centrifugal bulge passes over, the same area has its second high tide for that day. Thus, the two tidal bulges continuously travel around the earth about twelve hours apart. The two low, or ebb, tides take place midway between the two bulges. Since it takes twenty-four hours for the earth to revolve on its axis, it would seem these four tides would be timed exactly six hours apart. This assumption is not true, however, because it does not take into account the fact that the moon is continuously moving forward as it makes its twenty-eight-day revolution around the earth. This forward motion of the moon makes it necessary to add fifty-two minutes to each tide day.

For example, let us consider only the two high tides—the one when the moon is exactly overhead, and the other when it is exactly above the opposite side of the earth (Plate 3). Suppose the first comes at 10:00 in the morning. The next will come, not at 10:00, but at 10:26 in the evening. Then the next will come at 10:52 the next morning, not again at 10:00 o'clock as it had the morning before.

The sun has its own gravitational pull, independent of the moon. Whereas it is much larger than the moon, it is also much farther away from the earth, which makes the solar tide-producing force only three-sevenths of the lunar tide-producing force. As a result, it is the moon that has the greatest effect on the tides. The effect of the sun either adds three-sevenths to the moon's pull or subtracts three-sevenths from it, depending upon the relative position of earth, sun, and moon.

For example, extra-high tides, called spring tides, occur twice each month at the times of the new and of the full moon, when the sun, moon, and earth are in a straight line and the oceans are subject to the combined gravitational pull of the sun and the moon.

As the days of the month pass, the sun, moon, and earth change their relative positions, causing the daily high tides to decrease in intensity, until the sun and moon are at the corners of a huge right-angled triangle. When this happens, the solar and lunar pulls partly cancel each other out and we have unusually small high tides called neap tides. This occurs during the first and third quarters of the moon.

Unlike waves, which disturb only the surface of the ocean, tides move the entire body of water, and not just the water of the ocean. Freshwater responds to this pull just as seawater does, but the average lake or river is too small for the tides to be noticeable. As a matter of fact, the entire earth and surrounding atmosphere are moved by the sun and moon's tide-pulling force. Every time there is a vertical tide of ten feet the continents rise about six inches and the atmosphere bulges out many miles toward the sun and moon. We humans contain much

Spring tides with sun and moon pulling in a straight line below:

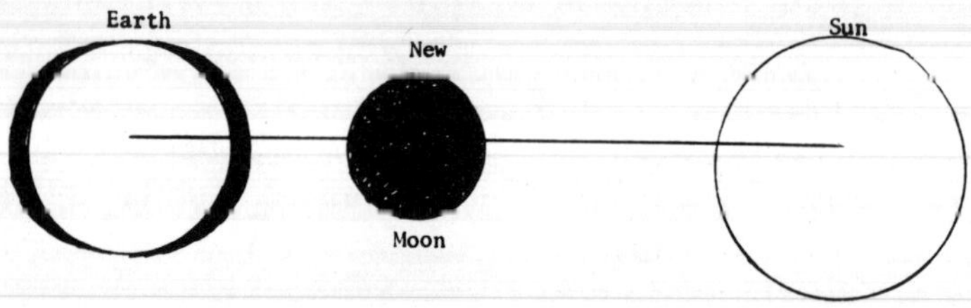

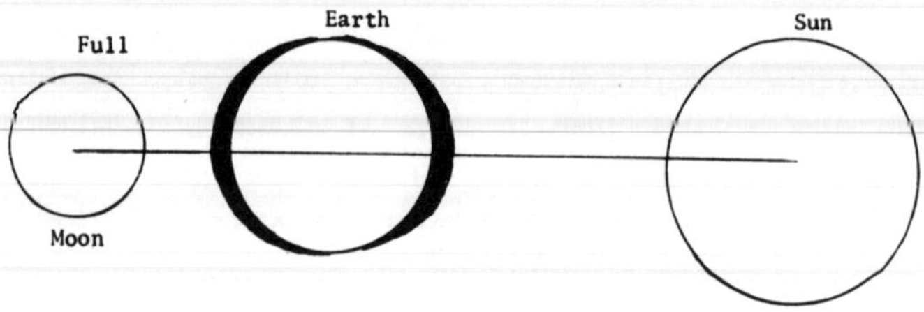

Neap tides with sun and moon pulling at right angles below:

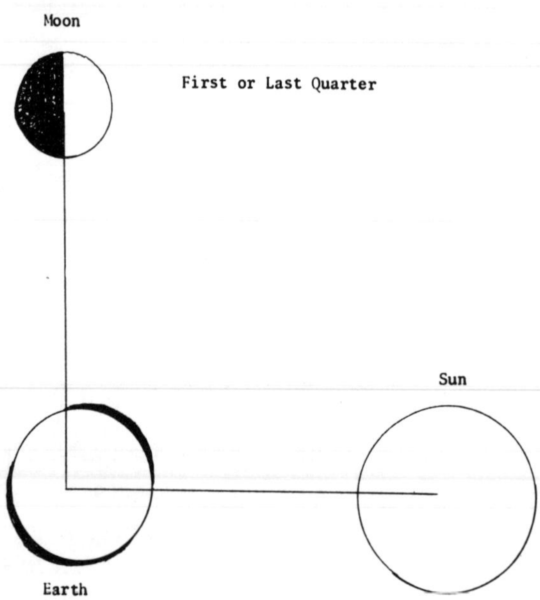

water and each of us is also subject to this tidal pull; we actually gain and lose a fraction of an ounce in weight with each rise and fall of the tide.

The difference in level between high and low water is known as *amplitude*. In the open ocean the amplitude is only two or three feet, so that ocean islands such as Tahiti and Hawaii experience a very small tidal range. Along the coast, however, the rise and fall is often far more. In broad bays with narrow openings the variation between high and low tides is often twenty or thirty feet. At each ebb tide, the level of the ocean lowers before the tidewaters of the bay have time to flow out. The bay entrance then becomes a raging torrent as the higher waters rush out to the lower level, and can present a great hazard to boats and swimmers. In one bay of this type, the Bay of Fundy in Canada, the tides can exceed fifty feet. Some one hundred billion tons of water move in and out of the Bay of Fundy with each tide change. The Mississippi River, by comparison, would need approximately 140 days to move the daily flow of this one bay.

The oceans do not cover the whole globe evenly, but are broken up into many different shaped basins of varying depths. Because of this, the water in each ocean basin surges back and forth in different ways in response to the tidal pull. Further complications are caused by weather. Onshore winds build up tide levels, whereas offshore winds lower them. Low barometric pressures increase amplitudes. Storm surges are produced when strong onshore winds accompany a rising tide. Finally, there are local variations due to the size of the continental shelf and the type of coastline.

Tides often run up broad open rivers such as the Amazon, the Saint Lawrence, and many smaller streams. The tide runs up the Saint Lawrence for 283 miles. In the Hudson River a two-foot tide extends 150 miles inland. In a few well-known rivers the tide advances upstream in a tidal bore. The bore is caused by a sandbar at the mouth of the river, which blocks the rising tide from entering the riverbed. The water builds up until it breaks through and starts racing upstream in a single wave, which bores its way against the natural flow of the river. Most tidal bores are harmless, with the boring waterfront no more than five feet high and moving forward eight to ten miles an hour. But some are much higher and are extremely dangerous. Tide bore waves as high as twenty-five feet go rushing up the Tsientang River in China. Behind the solid front of a tide boring wave, the rising tide gradually and slowly increases the depth of the river by as much as fifty feet, by the time it is fully in.

The Severn and Trent rivers in England, the Seine in France, the Amazon in Brazil, and the Petitcodiac in Canada are also known for their tidal bores. The latter is perhaps the most noted of all, as the Petitcodiac empties into the Bay of Fundy and must cope with the huge, sometimes fifty-foot tides for which that bay is noted. The tidal bore in that river is one of the major attractions of the North American continent and is visited by thousands of tourists each year. At Moncton, New Brunswick, a platform and parking lot have been built at a strategic high point overlooking the river where large groups congregate to watch the tidal bore go by below.

Engineers are now developing methods to make use of the vast source of tidal waterpower that has gone to waste for centuries. For many years they have harnessed our waterfalls, using the waterpower to produce electricity. But only recently have they recognized the rushing flow of the tides as a potential source of power. In 1966 France constructed the first successful tidal-electric plant near Saint-Malo above the mouth of the Rance River, utilizing a high, thirty-seven-foot tide range. A novel submarine-shaped generator was designed to utilize both the incoming and outgoing tidal flow.

Waves

True Waves

Try, if you can, to envision our globe in your mind's eye, and concentrate on the vast expanses of ocean that cover the earth. Then carry that vision a little bit further: picture the ocean waters—*all* the ocean waters—not as placid, smooth surfaces, but as covered with unceasing rhythmically rocking waves. These are the "true waves," which surge in constant motion over the entire ocean surface of the world. The alternate rise and fall of successive ridges of water are caused by friction between wind and water, and by other physical forces that are too complex for discussion here. Unlike the tides, the pull of the moon plays no part in the formation of waves, and the wave motion seldom causes water disturbance lower than thirty feet, as opposed to the tide disturbance of the entire water basin.

To anyone watching the waves from the beach, the whole ocean seems to be rolling in his direction as the waves travel shoreward, but that is merely an illusion. Actually, there is no forward movement of the water. Each particle of water composing a wave describes a curve and comes back practically to the point from which it started. Cork floats show that waves travel forward, but that the water itself does not. The cork moves up the front slope of the approaching wave, then slides down the back. And, when the wave moves on, the cork will not have moved more than an inch or two. So the motion of the ocean wave is somewhat like that of a waving field of grain, where the base of each moving stem is attached to the ground, though wave after wave passes over the field.

When a wave approaches a shoreline it begins to feel the drag of the bottom. This slows the lower part so the surface part moves forward more swiftly. The back of the wave now travels faster than the front so it rises to a peak and leans forward. Finally, the wave leans too far forward and breaks, thus ending its life at sea.

Wind Waves

With a picture firmly in mind of the "true waves" that surge rhythmically over all the oceans of the world, let us turn now to the entirely different and more localized wave group called wind waves. They do not simultaneously cover the entire surface of all oceans, nor are they constant; a wind wave exists while wind blows across a certain stretch of water, and as soon as the wind stops, the wind wave dies too. Just as sound waves range over a whole spectrum from low frequency to the highest pitch audible only to dogs, so do wind waves cover a wide spectrum of their own. Members of the wind wave family range in length between crests from a finger's width to thousands of miles, with the size of the wave in direct proportion to the strength of the wind.

The smallest waves, called capillary waves, are driven by a gentle wind, travel very slowly, and disappear as soon as the wind stops. Under pressure of strong winds, the waves become higher. They lean forward and break just as the smaller, constant true waves described above. However, the stronger wind forces the normal toppling action to be even more vigorous, so that when these heavy waves break, they produce foam—the familiar whitecaps. These larger wind-driven waves are called breakers, and they often cause tremendous damage.

Heavy waves can extend in size from the familiar choppy seas to the huge typhoon-driven waves. Extreme storm waves have been known to reach a height of a hundred feet. The surf of broken waves is sometimes thrown to heights of three hundred feet, with sufficient force to destroy lighthouses and move two-thousand-ton masses of masonry for considerable distances.

Waves traveling at different speeds may be whipped together by a storm, and combined into superwaves. The highest superwave so far observed, 112 feet, was encountered February 7, 1933 by a United States navy tanker en route from Manila to San Diego.

Other Waves

Wind waves are only a part of the whole wave family. Different types of water disturbance will produce different types of waves, but perhaps the "pebble wave" is the easiest kind to understand, because it can be seen in its entirety. Drop a pebble in a pond, and the resulting circular wave will travel across the entire pond to the bank. If a circular wave is caused by something small like a pebble, the resulting ripples are no larger than the small wind-produced capillary waves. However, if the pebble—the central point of force—should be an undersea explosion, such as a volcanic eruption or earthquake, the most fearsome, destructive wave of all, the tsunami, is formed. It surges out and fiercely spreads in all directions for thousands of miles across the sea, just as the tiny ripple disturbance crosses the pond. The huge tsunami (Japanese for harbor wave) is often called, incorrectly, a tidal wave, and has been known to travel eight thousand miles at a speed of five hundred miles per hour, and strike a coast with a fifty-foot wall of water. When the volcano Krakatoa blew up in 1883, 135-foot waves reached Java, killing 36,000 people.

It is now believed that the appalling loss of life caused by the tsunami may soon be a thing of the past. With modern technology, scientists are now establishing warning systems. In 1946 the United States Coast and Geodetic Survey set up a network to flash warnings of undersea earthquakes that might start the huge waves rushing across the Pacific. The warning traveled fast enough to save many lives in the 1960 Chilean earthquake, but many other lives were lost when warnings came too late. The networks are now being expanded for the protection of many areas in the hope of eliminating future tsunami death tragedies almost entirely.

Another type of wave, called the internal wave, occurs below the sea surface, at the boundary between water layers of different temperatures. A different type, the standing wave, moves up and down but does not travel. They occur in landlocked bays and harbors.

Sometimes a large wave, looking more like a big bulge, will seem to travel across the ocean surface without breaking. It is called a swell.

Great freak waves, called surges of the sea, or storm surges, sometimes occur under unusual meteorological conditions. These large masses of water, which surge in a violent, rolling motion, often cause great destruction.

Wave Terms

The top of a wave is its crest. The depression between two crests is the trough. The horizontal distance between two crests is the length of the wave. The vertical distance from the bottom of the trough to the crest is the height. And the time required for the crest to travel the length of the wave is called the period of the wave.

The size of a wave is determined not only by the strength of the wind, but also by the fetch, the extent of open water across which a wind can blow. When a wave comes from the open sea into water so shallow that it drags bottom, the crest pitches forward as surf.

The water forced against the coast and onto a beach by a wave runs back toward the sea again, and this away-from-shore motion is called undertow.

Whenever waves strike the shore at an angle, the water moves more or less along shore, producing a shore, or littoral, current.

Shorelines

Beaches

Down through the centuries the ocean waves have worked continuously each minute of every day and night to build and reshape shorelines. They wear down rocks into pebbles, and then grind the pebbles into sand. It can take several hundred years of constant motion to grind a one-foot rock into sand. Some sands are soft and grind down to mud, which is then carried out to sea and deposited on the ocean floor as layers of sediment. Others are hard quartz crystals that compose most of the sandy beaches on earth.

It is these tiny grains of crystal sand carried by the waves that shape most of the world's coastlines. They act as sandpaper in constant motion, always moving and cutting to eat away the softer sandstones along the shoreline, leaving the harder rocks as outcrops, or stacks. Our beautiful Oregon coast is an excellent example of such sand carving.

As the current moves along a coast, it often deposits its load of sand, forming long beaches. The texture of the sand determines the kind of beach the sea will make. Coarse sand is like blotting paper: waves sink into it, deposit their load, and it piles up loosely. The resulting beach is steep and constantly moving around. Fine sand packs much more tightly. Normal waves do not sink into a beach made of fine sand. Instead, their action leaves it smooth, hard, and with a gentle slope. However, even fine sand beaches are penetrated by storm waves. It is amazing how quickly a beautiful, long beach can be entirely washed away by the action of hurricane waves. One day the beach can be edged by a row of lovely beach homes, and the next day some of those beautiful homes will have toppled over into deep water where the beach used to be; and often other homes, not quite so undermined, are left to balance precariously at the top of a high sandbank just ready to fall (Plate 4a).

When such destruction happens, there is, of course, tremendous loss of shell life. Whole colonies of shells are washed out to sea with the sand, and only a small percentage survive to find a new hospitable environment.

Besides hurricanes, other tremendous forces of nature make radical changes in shorelines, and in the kinds of life that can inhabit them. As the centuries pass, some areas of the earth have a tendency to rise higher above sea level than formerly, and other areas have a tendency to sink lower. When a coastal area rises higher, new lagoons, deltas, sandbars, and sand dunes begin to emerge from the sea and form an emergent coastline. The reverse of this process, a sinking lower into the sea, causes new bays, inlets, and offshore islands to appear in what is called a sunken, or submergent, coastline.

Sometimes a river valley becomes submerged several miles back from its opening on the sea, a phenomenon called a ria shore.

Fjords

Another coastal variation is caused by huge glaciers grinding their way down the sides of coastal mountains to reach the sea. Then, perhaps centuries later, climatic conditions change so that the glaciers melt. With the ice gone, the coastal mountains are left with big empty gashes, which fill with ocean water. The gashes, called fjords, are a passage through which ocean vessels can travel many miles inland. Frequently the fjords are barely wide enough for a boat to squeeze through between the perpendicular mountainsides, often thousands of feet high. Such a fjord coast, with its series of gashes, is one of the most beautiful sights on earth. One important fjord region is along the coast of Norway, and another lies by the west coast of New Zealand's South Island. I can personally vouch for their grandeur, and for the fact that no other type of coastline approaches their beauty.

Estuaries

For a more complete understanding of the ocean coastal line, mention must be made of the estuaries that are so vital to the life of many ocean creatures. The name *estuaries* is used for the often broad and brackish areas where freshwater and ocean waters mingle.

Surprising as it may seem, there are 27 million acres of estuarine waters and wetlands in just the twenty-three coastal states and the territories of the United States. The largest and most significant are located in the vicinities of San Francisco Bay, the Mississippi Delta, the Florida Everglades, and Chesapeake Bay. One-third of our nation's population now lives in the 258 counties that border estuarine waters. And about one-third of our rapidly expanding industries are now concentrated there.

Scientists are gravely concerned lest rapidly encroaching civilization unwittingly and irrevocably destroy the estuaries until they can no longer perform their vital role in the perpetuation of sea life. They serve as a spawning ground, nursery, feeding area, and passageway for a great variety of aquatic life (Plate 1b). Marine biologists estimate that perhaps as much as three-fourths of our commercially important seafood—clams, fish, oysters, scallops, crabs, and lobsters—depends upon the estuarine waters for part or all of its life cycle.

Estuaries also provide many recreational opportunities for people, as well as providing food and shelter for millions of birds. They also serve as buffers against the ravages of violent storms. Yet pollution and landfills continue to destroy them at an amazingly rapid rate as "civilization" advances "forward." In California alone, 255,800 of their 381,900 estuary acres were already destroyed at the time of the 1966 Fish and Wildlife Survey. Pollution of Boston harbor has forced the closing of one thousand acres of shellfish beds at an estimated annual loss of $1,400,000. The Sports Fishing Institute estimates that estuary destruction in Galveston Bay is causing an annual loss of $25,622,000.

The unattractive appearance of these smelly and mucky salt marshes belies their importance in the total scheme of things. They are totally lacking in the type of majestic beauty that inspires nature lovers to fight in their defense. Thus, more and more of them are either being filled in for development, or else being turned into open sewers as they carry away trash, oily waste, and toxic pesticides and herbicides.

This pollution kills not only aquatic animal life, but plant life as well. Estuaries are among the world's most fertile areas, the richest producing ten tons of plants per acre each year. One marsh plant called cordgrass produces seven times as much food value per acre as wheat. As the marsh plants decompose, their nutrients are, by tidal action, carried into the water of the oceans to nourish marine life. Some marine scientists believe that this shallow water at the edge of the sea is by far the vital food-producing part of the sea, and that the deep waters are basically nonproductive. Thus, it is believed that further destruction of the estuaries will eventually adversely affect all ocean life.

The Federal Water Pollution Control Administration is now very much alert to the threatening danger and is moving in the direction of legal action to halt, or reverse, present trends. The American Littoral Society has also begun a campaign to help. Concentrating on the area from Maine to the Gulf of Mexico, it plans to take a complete inventory of estuarine resources, conduct a conservation education program, and work with the government to legally enforce some kind of proper balance between human and wildlife needs in the estuarine areas. At present, only time can tell whether help comes too little and too late.

2. The "Tree of Life"

Even more interesting than the ocean itself is the life it sustains—both plant and animal life, and especially the interrelationships between the two (Plate 5a). The physical creation of the universe is surely a great mystery, but the mystery surrounding the origin of life is even more profound. After centuries of skillful probing into the mysteries of the living cell, we are still, in the twentieth century, left with two schools of thought.

According to one school, the first life began purely by the chemical reaction of inanimate substances in the "primordial soup" that was then the ocean. From that single cell beginning, higher and higher forms evolved in the waters of the ocean until many crawled out on land. Then, through still more evolutionary changes, they became land animals and plants.

According to the second school of thought, an Almighty power had some part to play in the creation of life. To satisfy the second group, many questions still need to be answered: If made purely by chemical reaction, just how, or why, did that first animal live, breathe, and reproduce? No one even faintly questions the observable and provable fact that all living things evolve from one form to another as they adapt to changing circumstances. But what put the breath of life into the first one of each kind? What gave it the ability to adapt to changing circumstances? Why can't a living, breathing, functioning organism capable of reproduction be produced by the union of inanimate chemicals in a laboratory test tube? Scientists can make a perfect-looking egg in a laboratory, but they cannot make it hatch. Why not, if life is a mere chemical matter? How are the interrelationships such as that between a flower and a bee explained? Back in the beginning when the bee was not yet a bee and the flower was not yet a flower, how did they reproduce and keep on evolving when neither could have given its life-sustaining service to the other? Did they just suddenly and simultaneously, at some stage of their evolution, become a bee and a flower, with a vital need for each other that had to be met before either could produce even one more generation?

Believing that all living things evolved from a chemical union of nonliving elements, the scientist probes ever and ever deeper into the living cell trying to prove his theory of inanimate beginnings. On the other hand, the naturalist, who believes that an Almighty Power we call God instigated such miracles, feels profound confidence in his belief. How can he doubt, as he observes thousands of animals, plants, and insects, each giving their life-sustaining service to the other, while they all work and live in harmony with the revolving planets through daily, monthly, and yearly changes of tide, temperature, moisture, darkness, and light.

Few mysteries in nature are as baffling as the biological time clock built within the bodies of all living things, up to and including man. In the ocean, to take one instance, animals living between tide lines must adapt to alternating conditions of submergence and drying out. The change in tidal range from high to low is very vital to them, so nature, with wisdom beyond all understanding, has built a tidal rhythm within her ocean creatures. By instinct they are able to incorporate that ever-changing tide into their life cycle and habits. Some breed only at certain stages of the tide. Many gauge the timing of their egg-laying entirely by the tides. And the feeding patterns of all beach-type mollusks must be built around and conform to the coming and going of the tide. Some must wait until the tide comes in to eat, and others must wait until it goes out. But with the guiding hand of nature, all are able to survive a most difficult situation.

The overall plan of that total coordination seems too fantastically executed for the naturalist to doubt that it is the creation of an Almighty Power. And the deeper the scientist probes into the structures and functions of the protoplasms, enzymes, and hormones of the living cell, and the further he reaches back into eons of time, through the study of fossils, to prove his chemical-union theory of the beginning of life, the more he seems to prove the naturalist's theory of creation. The more marvels he reveals with that probing microscope, the stronger becomes the conviction of the naturalist that nothing short of an Almighty Power could execute such precision of interdependent activities between the body parts of plants and animals, and then endow them with the ability to reproduce and adapt to changing conditions down through the centuries.

The above questions and doubts are but a tiny few of the literally thousands of things in need of clarification before both schools of thought merge into one as facts replace theories in the minds of each. Meanwhile, let us proceed with our discussion of the plants and animals of the ocean as we know them to exist today.

Plankton

Ocean plants have the ability to capture energy from sunlight and manufacture their own food from the chemicals dissolved in the sea. With the chemical reaction of carbon dioxide in seawater and sunlight, the plants produce carbohydrate foods and release oxygen. By so doing, they make ocean animal life possible. They not only provide the vital oxygen that ocean animals require, but the plant itself provides food for the animal, which is unable to manufacture its own.

The one-celled plants and animals at the very bottom of the "tree of ocean life" are called plankton. Plankton includes bacteria, algae, the larvae, or very young, of any sea animal, many minute crustaceans—crab- and spider-type animals—and thousands of single-cell animals called protozoans and rotifers. Plankton is found floating and drifting in all ocean waters, as well as in all bodies of freshwater.

Surprisingly, plankton life thrives more abundantly in the cold Arctic and Antarctic areas than in the central equatorial areas. But, in every area, it is to the sea what grass is to the land: the basic source of food for all animals. Just as the huge buffalo can thrive on tiny blades of grass, so can the largest known whale thrive on the even tinier plankton plants and animals of the sea. The ocean tree of life builds up from small to large exactly as does the land tree of life with which we are more familiar. The tiny single-cell animals feed on the tiny single-cell plants, and of the meat-eating animal species, the larger feed on the smaller.

The plankton plants, called diatoms, supply nine-tenths of the food of the ocean. They are far more common and numerous than the plankton animals that are called copepods (Plate 5b).

One authority estimates that 10,000 pounds of diatoms eaten makes 1,000 pounds of copepods; 1,000 pounds of copepods eaten makes 100 pounds of smelt; 100 pounds of smelt eaten makes 10 pounds of mackerel; 10 pounds of mackerel eaten makes 1 pound of tuna; and 1 pound of tuna eaten makes .1 pound of man.

Many researchers believe that the minute plankton and algae plants of the sea will eventually play a major role in solving the coming world food shortage. Scientists think that this primitive form of leafless, stemless plant life will someday feed crews in spaceships, and even appear on supermarket shelves to be eaten directly as a major food item.

Indiana University is cultivating and studying 1,500 strains of algae in one of America's most compact test-tube "farms." The shapeless patches of brown, green, and pink minute plant growths, in their multitude of test tubes, are grown in rooms with controlled light and temperature conditions. Dr. Richard C. Starr started this "farm" in 1953, and it is now one of the world's major research centers for the interesting and perhaps far-reaching study. Other research is being done in Czechoslovakia, at Cambridge, England, and at the University of Gottingen in Germany. Dr. Starr is also helping start a collection for research at the University of Tokyo in Japan.

The amazing speed with which the single-cell ocean plants multiply makes for rapid advance in their study. In some strains, a single cell literally explodes into 65,000 cells within a single twenty-four-hour period. Thus, it is possible for science to complete a study of thousands of generations in the time needed for the same study of but a few generations of slow-growing plants or animals.

Though plankton is the extremely valuable life-giving force of the sea, it occasionally causes a very destructive phenomenon known as the red tide. The red tide of Florida and the Gulf of Mexico is caused by the "bloom" of a microscopic plankton called *Gymnodinium breve*, which is always present in the coastal waters of western Florida. Under normal conditions, no more than a thousand of these plankton will be found in a quart of water. But during this intensive blooming, a quart of water can contain as many as sixty million.

During the bloom periods, when the sheer quantity of plankton turns the water rosy—whence the name *red tide*—it reduces the oxygen content of the water, and also gives off an irritating gas when dying and decaying.

A similar red tide, caused by a related organism, has occurred off the California coast. Yellow-water blooms have been reported off Vancouver Island, British Columbia. In the Indian and West Pacific oceans, red tides also occur.

A phenomenon similar to the red tide occurs periodically on the Great Barrier Reef along the Queensland coast of Australia. The local people explain it by saying the reef is "cleaning itself." Normally, the water on the reef is amazingly clear. You can look into it to surprising depths. It shows beautiful blue green colors as you look out over the reef area for miles beyond the coast. But all that changes while it "cleans itself." The water then assumes a dirty, brackish appearance as far as you can see. I watched it happen, but was unable to locate an explanation.

Sponge

The peculiar "animal" that we call sponge is classified next above the single-cell (protozoa) animals of the plankton.

Over three thousand sponge species are known to exist in the waters of the world. For centuries there was much controversy over the sponge's classification. At one time it was believed to be a plant. At another time it was thought to be a nonliving substance secreted by the many animals that take refuge in its cavities. Not until about a hundred years ago was it fully determined to be a living animal. Because of its millions of pores, the sponge was given the phylum classification name Porifera, meaning pore-bearers, and it is the "lowest," or least complex, of the multicellular animals.

The body mass of the sponge animal (Figure 1) is penetrated by many channels, some of which are inlet tubes, some outlet tubes, and others connecting tubes. The outside is covered with a thin leathery membrane called the ectoderm. The tubes are lined with a softer membrane, the endoderm. Between these membranes is the mesoderm. This middle layer is strengthened by fibrous, limy, or glassy material, and actually constitutes the sponge as we know it in our homes.

The tube linings are all covered by gently moving cilia (or hairs) that create a current to draw the seawater into the entrance tubes, force it through the connecting ones, and on out the exit tubes.

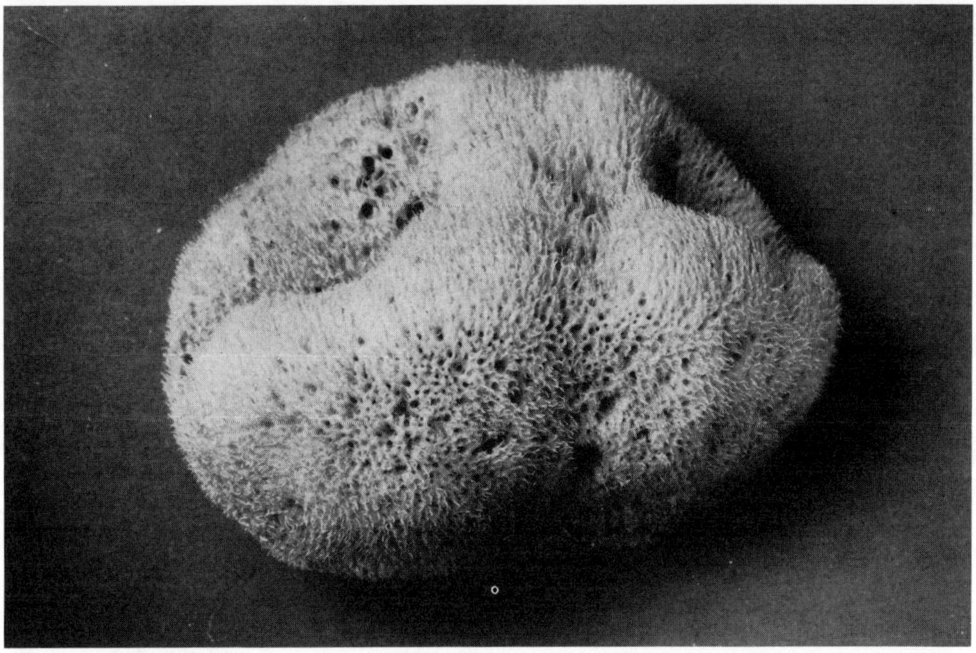

Figure 1. A turkey cup sponge from the Mediterranean Sea (5 inches across). *Courtesy U.S. Fish and Wildlife Service*

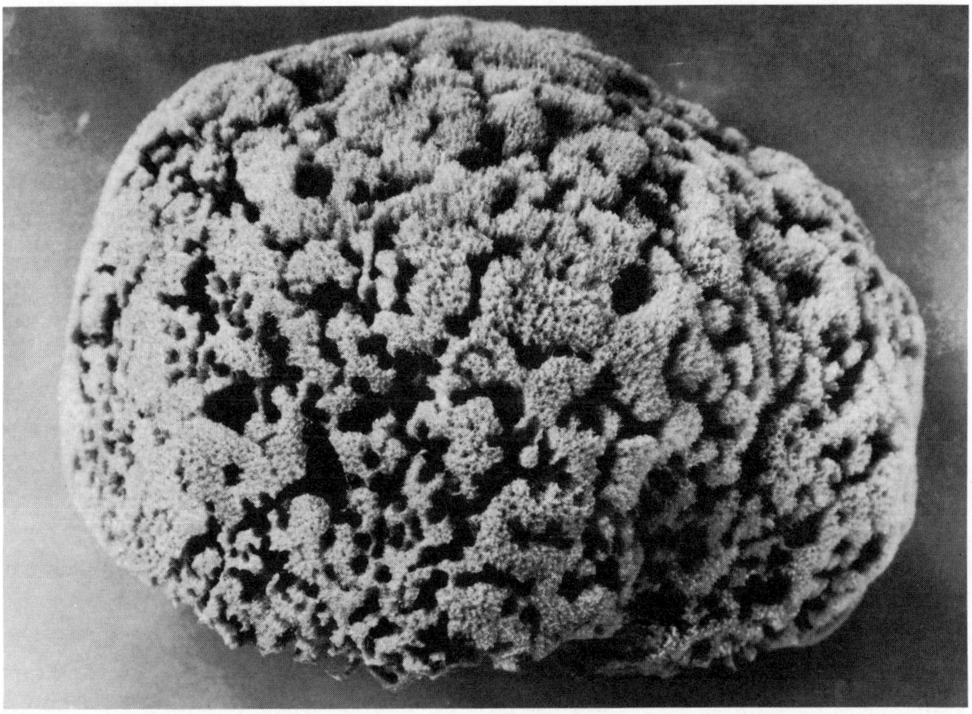

Figure 2. A Florida yellow sponge (8 inches across). *Courtesy U.S. Fish and Wildlife Service*

As the water passes through, the plankton food is strained from the water and digested by tiny cells lining the tubes.

The tube structure of the sponge provides the perfect home for a large number of other animals. Small shrimp, marine worms, small crabs, young larval forms of sea life, small snails, clams, and many other animals live within the channel tubes or lie partially concealed in the body of the sponge, but they do very little damage to the sponge itself. Many of these animals live on the small food particles in the water circulating within the sponge.

Some species of sponge reproduce sexually, by fertilized eggs, and other species reproduce asexually, by budding and branching. Of those produced by eggs, in some species both male and female cells arise in the same individual. In other species they occur in different individuals, and the male sperm is brought into the female sponge with its water currents. When the eggs hatch, the infant larva for a time moves about freely in the water. The tiny infant is covered with moving hairs that propel it through the water. A little later, it settles and attaches to a solid surface where it will stay for the remainder of its life. It then turns inside out and the hairs that had been moving it about become the tube hairs that start the circulation of water through the animal.

The sponge species that reproduce by branching send horizontal branches out over the rocks from which new growths arise. In other species buds, containing the nucleus for growth, break away to grow into new, separate, and complete individuals.

The sponge has a remarkably unique ability to regenerate a new individual from broken parts. Some species can be forced through a fine sieve. Then the dusty residue, if placed in a pan of sea water, will regenerate into a whole new sponge.

Few people are aware of the colorful beauty of living sponges. They constitute one of the most beautiful "flower beds" to be found anywhere in nature. Unfortunately, no way has been found to preserve the color when they are removed from the water. Seeing only the drab color they fade to when dried, people just assume that, in life, they are no different in color from the familiar bath sponge.

Often during storms, much sponge growth is broken off and carried ashore by waves. It was my good fortune to be collecting in one area when this happened. It was the most thrilling and colorful beach sight I ever encountered. In a one-mile strip of beach I collected thirty-five different species representing almost every shade of the color spectrum. Although it was impossible to preserve the color, it makes a valuable educational addition to my ocean-life exhibit. Most people, being familiar with only the household sponge, have much to learn from just the wide variety in structure in this exhibit.

As an object of utility, the sponge has a long and distinguished history. According to the early writings of Plato and Aristotle, sponges were supposed to have been first used by the Phoenicians and Egyptians. They were so popular among the ancient Greeks as padding for helmets and leg armor that the men who cut great numbers of them from the ocean floor developed superior diving skill, which led to the introduction of diving as part of the original Greek Olympic games.

This Greek diving tradition had an interesting extension into modern times. The Greek community at Tarpon Springs, Florida, developed a great sponge-fishing industry there. Their great diving ability enabled them to take advantage of the fact that, of the thirteen commercially useful sponge species, eight thrive in the Tarpon Springs area.

The divers work down to a 120-foot depth, and carefully remove the sponges from the rocks to which they grow attached. They are careful to leave the base in place so it will grow into another sponge, and thus produce repeated crops. After being brought to the surface, they are cleaned of the living animal matter, trimmed, and then sold through the Tarpon Springs Sponge Exchange.

When the industry was in its prime, thousands of tons were harvested yearly. For many years great quantities were shipped to points throughout the world. Then because of sponge disease and overcollecting, the industry began to rapidly decline. The natural sponge became so costly it could no longer compete in price with the artificial cellulose sponge, which has now largely taken over the sponge market. The natural sponge is, however, still in great demand by glass factories. Since it will not burn, it is ideal for wiping hot glass.

The natural sponge is still of enough value that considerable effort is now being made to propagate it artificially. The living sponge is cut into small pieces and the cuttings are carefully placed in favorable localities. The method has already been proven practicable; pieces planted in Florida waters grow to a marketable size in one year.

The revival of the industry has become increasingly important as medical research progresses. It is now known that a few sponge species have poisons in their bodies that seem to hold great promise for curing certain types of diseases. The red-beard sponge, for example, contains a poison called ectyonin that is capable of killing a wide variety of microbes and viruses, thus making it an effective anti-infection agent. It even destroys the fierce staphylococcus bacteria that penicillin is powerless to conquer, and it may eventually become an effective weapon against tuberculosis, trench mouth, and certain bladder disorders.

This general discussion of the sponge is not complete without mention of a few special species of outstanding interest. Perhaps the one most remembered by most people, and remembered with the highest degree of mixed emotions, is the fire sponge, a beautiful bright red sponge commonly found on coral reefs. It is a gorgeous thing to behold, but woe unto the uninformed who does more than merely behold. This species can severely paralyze fingers and cause great pain. And, just as with a similar injury from poison ivy, considerable time is needed for the swelling and pain to subside.

Another sponge of mixed blessings is the bright yellow boring sponge (*Cliona sulphurea*) (Figure 2). It is a fine-textured sponge that grows in large irregular masses, which performs a peculiar service to the economy of the sea from Cape

Cod to South Carolina. Its growth spreads over and completely envelops the surface of dead shells. And as it grows it penetrates and honeycombs the shell, until the shell is completely destroyed by absorption. Without this sponge, dead shells would possibly accumulate in disturbingly vast quantities. However, as it goes about its business of doing good, it does harm at the same time. Many times it will settle on a living shell and begin its penetrating, honeycombing procedure. The living animal then secretes lime so rapidly to repair and cover the perforations that it often exhausts itself and dies.

A third sponge, the most beautiful of all the sponges, must be explored here before we can rightfully close this section. It is popularly known as Venus's-flower-basket, or the watering-can sponge, but carries the tongue-torturing scientific name of *Euplectella aspergillum*.

Certainly nowhere in nature has a more beautiful, dainty, and intricate creation been perfected (Plate 5c and d). It has the appearance of gorgeous handmade lace woven with fine glass fibers no thicker than fine human hair. It is woven in the form of hollow tubes often shaped something like a bull's horn. A beautiful, sometimes lace-edged, cap covers the large end, and from the smaller tapered end extends a multitude of loose glasslike fibers that anchor it in the mud at the bottom of the sea. Because the fibers so strongly resemble fine spun glass, it is also often referred to as the glass sponge.

It grows at very great depths, from three thousand to fifteen thousand feet, and is found only in warm seas, usually in the area of Japan and the Philippine Islands. It is sold in Japanese curio stores, and people actually believe it is delicate handwork made by skilled oriental artists.

A curious Japanese tradition has grown up around Venus's-flower-basket. A tiny shrimp, genus *Spongicola*, lives in conjunction with Venus's-flower-basket. While young, the shrimps swim freely in and out of the meshes of the tube. But, as they grow, a pair of them will often become trapped and must spend the remainder of their lives inside the tube. The people of Japan are so intrigued by the happy togetherness of this tiny couple that they accept Venus's-flower-basket as a symbol of permanent married bliss. To ensure the happiness of a Japanese bride, it is customary to give her one of the baskets bearing its two tiny occupants.

Ocean Plants: Seaweed, or Kelp

As we travel up the tree of ocean life, we will now turn our attention to higher organisms of the ocean, the plants commonly known collectively as seaweed, or kelp. The lower plankton plants are noted for their fabulous numbers, while just the opposite is true for the more complex plants of the ocean. There are no more than one hundred species of the larger ocean plants, as compared to some 250,000 land plants. Although few in number of species, they grow profusely. They not only furnish a haven of refuge for many sea animals, but they are now used commercially in a surprising number of ways.

Seaweed is a very popular fertilizer,

Plate 1

One of the author's shell designs. *Robert Rickert*

Plate 2

A sandy beach during high and low tides.

Plate 4

(a) This beach, a victim of the 1968 Hurricane Gladys, has been so undermined it can no longer support the house.

(b) An estuary near Boca Ciega Bay, Florida.

Plate 5

(a) Some sea creatures, including starfish, Gorgonia, and soft coral.

(b) A microscopic picture of animal plankton (copepods). *Courtesy U.S. Fish and Wildlife Service.*

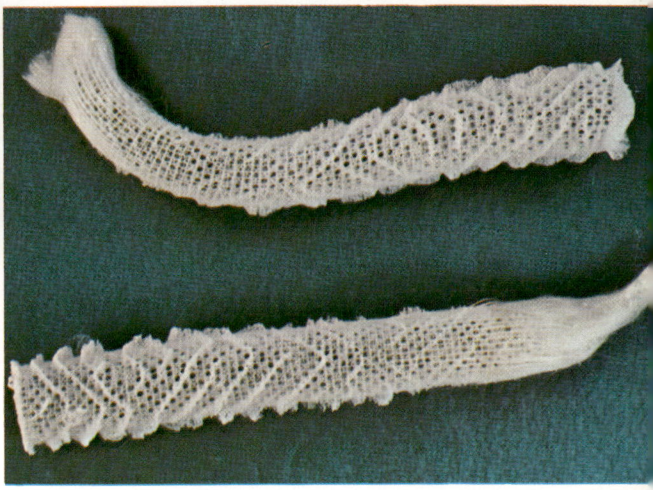

(c) Venus's-flower-basket (12 inches long).

(d) Close-up of Venus's-flower-basket filigree.

Plate 6

(a) Live Florida anemones, opened up for feeding (3 inches across).

(b) A 6-inch-across section of living coral taken at a depth of 60 feet from the Gulf of Mexico.

(c) A lovely "symphony in blue" of damsel fish swimming around a live head of coral. *Charles E. Herron*

Plate 7

(a) A coral atoll—one of the Kwajalein atoll chain—in the Marshall Islands. *Charles E. Herron*

(b) Sea urchins. *Bottom:* live sea urchins. *Top left:* urchin underside, cleaned of animal matter. *Top right:* topside of a cleaned urchin. *Center:* urchin's teeth.

(c) *Left:* topside of the starfish (4–5 inches across) common to the Tampa Bay area of Florida. *Right:* underside of the same species.

Plate 8

Topside of a live basket star (18–24 inches across) found at a 40-foot depth in the Gulf of Mexico. Its internal organs lie in the center, and all the arms and branches move continually. It is resting on *Gorgonia,* a branched, plantlike animal related to coral.

(a) Blue crab.

(b) A fiddler crab, close by the entrance to its burrow.

(c) A fish cleaning station. *Pete Hogg*

Plate 10

(a) A hermit crab "at home" in the empty shell of a Scotch Bonnet.

(b) Horseshoe crabs. The largest one is approximately a foot long. They are living fossils, closely related to the spiders.

(c) An empty whale barnacle colony. Each "cup," about 2 inches across, used to have a living animal inside.

(d) Crucifix of the Sea (4 inches long). The figure of Christ on the cross is startlingly clear on the left (*underside*), and a priestly figure seems etched in the skull of the right (*topside*) example.

Plate 11

(a) A roughtail stingray. *Courtesy U.S. Fish and Wildlife Service*

(b) *Top left:* a mature female sea horse. *Top right:* a male sea horse, approximately 5 inches long, showing the pouch bulge in which he carries the babies. *Top center:* baby sea horses. *Bottom:* a sea horse in the author's collection that tried to swallow its tail—possibly occurred during the preservation process. *Robert Rickert*

(c) A green turtle surfacing for a breath of air. These turtles weigh from 200 to 300 pounds each.

Plate 12

(a) Shore birds feeding on small animals and shells.

(b) A sea gull scaveng-ing on the beach after a storm.

(c) Large sea beans, measuring 2 inches across.

and when burned is a valuable source of iodine. Scientists believe it will soon be widely used in treating allergy patients, since its extract can control the release rate of injected drugs. This permits doctors to administer higher concentrations of drugs at one time, thus decreasing the number of shots needed. Seaweed extract is already an important ingredient in more than half the toothpaste produced in the United States.

A chemical called agar is produced from red seaweed. Each year a million pounds or more of agar are used in the United States—mostly by bakeries as an additive in cake icings, and as an activating ingredient in laxatives. Brown seaweeds contain alginic acids of wide medicinal use. One or another of those chemicals is an ingredient in almost all dietetic foods. Seaweed chemicals make puddings jell, milk drinks thicken, and cookies firm. They also hold various food elements of drinks in proper suspension, just as they keep various drugs suspended in liquid medicines.

A few of the ordinary foods in which we (usually unknowingly) eat seaweed extracts are: ice cream, canned milk shakes, infants' formulas, evaporated milk, milk puddings, no-bake custards, whipped desserts, pie fillings, cheesecake, instant party dips, frozen pies, jellied fruit juices for infants, jellies, soft drinks, and salad dressings. The Japanese are already quite advanced in their study of seaweed itself as a source of human protein food. And a seaweed extract called ebimer is helpful in the treatment of peptic ulcers.

At the beginning of the twentieth century, Louis Pasteur became aware that much sea life contains antibiotic substances, but the subject was not actively pursued until the past twenty years. Some 150 species of algae have been studied in Japan, Puerto Rico, California, and other areas. It is anticipated now that many human infections will be cured by antibiotics extracted from sea plants.

Acids derived from kelp are also effective in ridding the human body of the dangerous, radiation-bearing strontium 90 that comes from atomic fallout. Strontium 90 rains down on crops from the upper atmosphere, both cattle and people eat the crops, and then the strontium 90 enters man's system primarily by way of cereals, dairy products, and meats. Once eaten, it is absorbed into the bloodstream and finally deposited into the skeletal tissues where it can remain for years as a constant source of radiation affecting the bone marrow, and encouraging the development of bone tumors and leukemia.

In the last few years strontium 90 levels have diminished because there have been fewer nuclear bomb tests. But, in the event of future large-scale testing, or nuclear warfare, it is now believed that kelp could save many people from serious radiation-related illnesses.

Off the coast of California is the largest known giant kelp bed (*Macrocystis pyrifera*) in the world. It is called the sequoias of the ocean, and is rooted by sucker-type holdfasts in fifty to sixty feet of water, and the stems are heavily covered with leaves that frequently reach a length of 150 feet. They are buoyed up by rows of air-filled bladders attached at intervals along the stems.

To harvest seaweed, kelp barges go back and forth over the beds just as a grain combine travels over a wheat field. At the front of the barge a ten-foot bar cuts off the stems beneath the water surface. The harvested kelp is carried into the three-hundred-ton-capacity boats on an endless conveyor belt. Instead of destroying the bed and depleting the supply, if properly done, the harvest actually improves the bed. It removes the old shoots, thus giving the vigorous young growth a better chance to sprout and develop.

For many years there seemed little danger of ever exhausting the supply. Then, as so often happens, man upset the ecological balance of nature in such a way as to destroy much of the valuable kelp, with a serious threat facing what remains.

In the overall plan of nature, kelp is a natural food for the sea urchin (*Strongylocentrotus franciscanus*), and the sea urchin is a natural food for the sea otter. Man, from his huge cities and shore installations, dumped sewage into the ocean, which increased its nutrients, which in turn caused the urchin population to greatly multiply. At the same time he hunted the sea otter almost to extinction, so that fabulous numbers of urchins destroyed and devoured the kelp beds.

As so often happens, after the damage is done, foolish man learns that nature's plan is best after all. In an attempt to restore the balance of nature that had been so ignorantly destroyed, laws were passed to protect the sea otter. At the same time, great sums were spent on research aimed toward finding a way to artificially check the fabulous sea urchin population, and thus restore the extremely valuable kelp beds.

As of now, the results are both good and bad. The protected sea otters soon multiplied to a point of overpopulation in certain areas, so that they themselves began destroying other valuable life, such as abalones and so on, while at the same time, in some regions, such as Santa Barbara, they did reduce the incidence of sea urchins, making possible the natural reestablishment of some kelp beds.

Costly experiments still continue, with a few kelp beds brought under artificial cultivation with no help from the sea otter. Scientists have learned that it is possible to destroy the urchins in limited areas with the use of lime, and thus plant and successfully raise kelp on small commercial farms. The problem now, though, is that pouring quantities of lime into the ocean for the sole purpose of raising kelp may again destructively upset the balance of nature in ways that we cannot foresee.

3. Coelenterates: Jellyfish and More

Coelenterates, distinguished by an extremely simple body structure, are one of the most primitive examples of multicellular animal life while, at the same time, some of its representatives are among the best known sea creatures. Anyone who has ever stepped inadvertently on a jellyfish and felt its sting, or has become entangled in its tentacles while swimming, has known a coelenterate intimately. This phylum, which is known to be composed of at least ten thousand different species, also includes anemones, coral, and sea pansies.

The body of a coelenterate, formed like a tiny, boneless sac, is all stomach, with a single opening through which both food and body wastes pass. Most species are equipped with tiny stinging cells, and many live in colonies made up of a huge number of individuals.

In this chapter we will discuss a few of the best known and most interesting families.

Jellyfish

Over six thousand species of jellyfish are known to exist—mostly marine species and a few freshwater—and they vary greatly in size and coloring, but all species look just like a soft quivering mass of gelatin. Most have poison tentacles hanging down with which to sting, immobilize, and capture their food. The largest species grows to a width of eight feet and its tentacles reach over a hundred feet in length.

The Portuguese man-of-war is the most widely known species of jellyfish because it injures so many swimmers and consequently receives much publicity. Its beautiful twelve-inch-long purple, pink, and red float (or bladder) goes sailing along on top of the water with all its danger hidden underneath. A dense cluster of tentacles hangs as far as sixty feet below. Oddly, these tentacles are several kinds of organisms differing from each other in both appearance and function. Some are for feeding, some for stinging, and some for reproduction. Each of the separate organisms constitutes a part of the whole animal.

All jellyfish species bearing poison tentacles are a constant hazard to ocean bathers. If a swimmer becomes entangled in these tentacles, he is in for a truly agonizing time. They contain a powerful paralytic toxin very much like cobra venom, and are considered one of the most dangerous of all sea animals. Contacting but a single tentacle triggers the release of thousands of tiny poisonous darts that paralyze breathing muscles and often cause death. It is believed this toxin may be the real cause of many unexplained drownings. Often ocean-swimming beaches in resort areas are closed to swimmers if many jellyfish are known to be in the area. Great care must also be taken when walking barefoot on the beaches. At certain seasons many jellyfish are washed ashore and stranded on the beaches, still with the power to sting.

Various methods have been found to somewhat ease the intense pain of the jellyfish, or sea nettle, sting. Rubbing alcohol or methylated spirits are helpful. It also helps to freely apply shaving cream and shave the affected area. The application of meat tenderizer is sometimes used with good results. The object is to destroy or remove the stinging bodies injected into the skin by the jellyfish. They remain active and continue to pump poison into the victim for some time if not removed. The thing to carefully refrain from doing is to wash the affected area with water. Water only keeps the stinging bodies moist so that they can function longer, and the friction incident to the washing causes more of the bodies to discharge their poison. It has been learned in laboratory experiments that they often retain the power to sting long after becoming dried, as well as after being frozen. The sting cells easily penetrate clothing and, according to marine biologist Dr. Charles E. Lane, they can even penetrate rubber gloves.

Dr. Lane has spent fifteen years studying the man-of-war. He was the first marine biologist to isolate and analyze its toxin. It took him about five years of constant experimentation just to devise a system for obtaining pure venom from the creature. It is believed those gelatinous blobs may someday provide a cure for their own stings, plus a preventative for ulcers and a remedy for heart disease. The mere presence of a poison in nature alerts scientists to the possible existence of a curative drug. For substances that will kill creatures as big as human beings will often, when taken in minute doses, destroy harmful microbes that attack man, with no harm to the human.

Neither the breeding place nor the life history of the Portuguese man-of-war in the Atlantic is known. Extremely small specimens have been collected in the Miami, Florida, area, and are being studied at the Institute of Marine Science located there. It is presumed their breeding place may be somewhere near the latitude of Miami, although much study must yet be done and possibly many years may elapse before that carefully concealed secret is solved by man.

One of the most unusual and least

understood companionships exists between this poison jellyfish and the small blue- and silver-banded *Nomeus* fish. The man-of-war's poison tentacles bring death to the many small fish upon which the jellyfish feeds. Yet, to this special and beautiful fish, they serve as a haven of refuge. It is not understood why the man-of-war protects this one fish while consistently stinging everything else with which it comes into contact. As so often happens, this odd pair has established a kind of mutual-aid society whereby each contributes a valuable service to the other. The tiny bright fish darts about in the waters adjacent to the jellyfish to attract other fish. The larger fish, anxious for a meal, dart toward it. The tiny, and very clever, *Nomeus* quickly darts back into the poison tentacles to personal safety as it leads its pursuer to its death. Then both the jellyfish and the tiny fish share the feast.

The death-dealing man-of-war does have one mortal enemy. Once again the mystery of immunity is not understood, but the giant loggerhead sea turtles are unaffected by the man-of-war's poison. A huge five-hundred-pound turtle will swim through a bed of the jellyfish leaving utter destruction in its wake as it feasts without injury to itself.

The many other species of jellyfish, which are an important source of food for a large number of sea animals, are distributed in great numbers over a large area of the world. The largest concentration that it was my privilege to personally observe was in Puget Sound, while I was sailing among the islands off the coast of Washington. They were floating in the water by the thousands and were very easily scooped up with dip nets for close inspection.

Anemone

Though anemones resemble beautiful flowers, they belong to the coelenterate branch of the animal kingdom. There are a multitude of different species of varying shapes and colors. It is not known how they get their beautiful colors, since they often seem to live at three-hundred-foot depths where the sun's rays seem too diffused and filtered to produce color.

When closed and at rest, they appear to be just solid, unattractive bunches of flesh drawn into knots and attached to rocks or some other solid base. But, when they open up to feed, a broad high column rises up, and around its top hundreds of beautiful threadlike tentacles radiate like the petals of a large flower. They look for all the world like brightly colored chrysanthemums—any number of beautiful color shades all grouped together in one bed. Some of the roundheads are no larger than a teacup and others are as broad as a dinner plate.

Looking down through clear water at a fully opened bed of anemones is one of the truly beautiful sights of the ocean.

The stomach cavity occupies the whole interior of that raised column. The tentacles (or brightly colored "flower petals") slowly move about seizing any food contacting them and forcing it into the cavity where it is digested (Plate 6a). The indigestible portions are then thrown back out through that same mouth so well concealed in the center of that chrysanthemum-shaped top.

Anemones thrive well in saltwater

aquariums. However, the water in the tanks fails to supply enough natural food. Therefore, three times a week, the caretaker uses a long-handled tool and very carefully holds a small fish, mussel, or clam at the mouth opening in the center of each flower. It was my very great privilege to watch it being done at an Oregon aquarium. At the instant of awareness, those beautiful "flowers" began acting like the "animals" they were. The "petals" all started curling back to reach the food and force it into the stomach cavity below.

Anemones usually feed on the small animals of the sea, but they have the ability to capture and digest a fish of amazing size. Their tentacles are charged with thousands of nettle cells, or microscopic "harpoons," which eject a paralyzing poison and leave the prey helpless. One anemone, eight inches in diameter, was observed as it paralyzed and digested part of the thirty-inch shark with which it shared an aquarium tank.

Oddly, the anemone does not sting the human hand. It is believed the human skin is too thick for the tiny harpoons to penetrate. Even more oddly, there are six small fish species, all from one to three inches in size (with the scientific name *Ampiprion*), that seek shelter in the anemone and use those poison tentacles as a refuge.

The green anemone (*Cribrina*) grows profusely in tide pools. Its color is due to algae that actually live within the anemone's tissues. Another species, the *Adamsia,* is most unusual in that it attaches itself to shells being occupied by hermit crabs, and thus travels about by the hitchhike method.

Most anemones grasp a rock or some solid surface for support while reaching into the water for food. They appear to be permanently attached to their support, but actually they just grip it tightly with a suctionlike disk at the base and can slowly move about. Some deepwater species are unattached and swim about freely with a rhythmic muscular contraction movement. But the sand anemone burrows its tube into the sand, leaving only the flowerlike cluster of tentacles in the water above.

Some species reproduce by discharging eggs and sperm into the water. Larvae develop from the fertilized eggs and soon settle to the bottom to become baby anemones. From others, small fragments break away to become little anemones. And still others reproduce by dividing into two parts, with each part developing into a whole animal.

Sea anemones have never been known to suffer natural deaths, although they are frequently destroyed by other ocean animals and numerous unfavorable changes in environmental conditions. It seems almost too incredible to believe, but William F. Stephens, an authority, writes that they are known to have lived in captivity for a hundred years. When fully grown, some were placed in a tank at the University of Edinburgh in the year 1870, and they are still vigorously producing quantities of eggs and sperm each year.

It is quite impossible to remove an anemone from the water and preserve its beauty. Consequently, none can be shown in a museum case. What a pity. It is a real tragedy that such beauty can be enjoyed by comparatively few people.

Coral

Most people are familiar with the white coral on sale at gift shops. And some know the more rare red and black polished coral gems on sale in jewelry stores. Not so many people know, however, that coral is actually the skeleton of the tiny coral animal. They are found chiefly in the tropical Central Pacific, the Indian Ocean, the Mediterranean, and in the waters off the coast of Florida, Mexico, and the West Indies. The corals build colonies comprised of many individuals called polyps. Each colony has a shape characteristic of the species: branched, boulderlike, flat, and so on. The individual polyps are just small tubes attached at the base, with the open free end, or mouth, surrounded by beautifully colored tentacles. They nourish themselves by capturing in their colorful waving tentacles the tiny bits of food that float by, forcing the food down into the stomach-cavity tube for digestion. Some coral species can also sting their prey with a paralyzing poison.

All the polyps in a colony are derived from a single fertilized ovum that matured and then began to form new polyps by budding. As the individual polyps die, their flesh is washed away, leaving only the skeleton, which was made by the deposit of limy secretions from the mantle of the living polyp. The remaining living polyps stay attached to the skeleton, and in their turn produce buds. Thus, the coral formation with which we are familiar is actually the skeletons of many generations of the coral animals stacked one on top of another. Under favorable food and temperature conditions, they multiply and grow at a rapid enough rate to add each year a thickness of from one-half to three inches to the coral skeleton formation. The core becomes solid, with only the surface occupied by living polyps (Plate 6b).

The beautiful colors, which range all the way through the color spectrum, are intensified by an algae that lives within the soft polyp tissue. The algae absorbs carbon dioxide produced by the coral, and emits oxygen used by the coral, making each essential to the other's existence. So vivid and intense are the colors that a coral reef resembles a varicolored flower bed. Nowhere in my experience have I been dazzled by an array of color more startlingly beautiful than on the Great Barrier Reef (Plate 6c). In the warm equatorial waters off the coast of North Queensland, Australia, conditions are ideal for coral growth. Before the invasion of the crown of thorns starfish (see page 53), one could go out in a glass-bottom boat off little Green Island to enjoy nature's stupendous coral show. As one looked, and thrilled to the sight, how one longed to share the joy with everyone else. It seems such a tragedy that such glorious beauty has been seen and enjoyed by so few. But like the anemone, it is quite impossible, even where it still exists, to remove coral from the water and preserve its beautiful color for a museum case.

In most coral species the skeleton itself is white. There are only a few species with permanent color found in the skeleton. The beautiful Mediterranean rose coral is the most widely known of this type. It is solid, takes a high polish, and has been known to bring as much as $1,000 for one ounce. Also familiar is the red organ-pipe coral (*Tubipora musica*), which retains the color in the skeleton after the polyp that made it has died. Oddly, the living polyp that forms the red coral is green and not red as one would assume.

The third permanently colored coral skeleton, the blue coral (*Heliopora*), is a quite rare coral growing in a limited area off northeastern Australia. It is less well known in this country than most coral species. A number of years ago I was fortunate to obtain a piece brought back to the United States by a soldier who had been stationed in that area. Its lovely blue color shows only on the interior of the formation when it is broken apart.

A fourth coral species deserving of special note is the black coral (order Antipatharia). It grows in the form of small underwater trees, from rather shallow water down to great depths. Before my arrival in the Hawaiian Islands in January 1964, divers had brought up some magnificent specimens. They were carving beautiful and very valuable jet black statues and ornaments from the trunks of larger trees. I shipped home for my collection one of the smaller trees, which I treasure very highly.

Until recent years, black coral was believed to grow only in deep waters off the Hawaiian Islands. Very recently a professional diver found it growing rather profusely at Pandora Reef, off the coast of New Queensland, Australia. It has also been discovered in the Gulf of Aden, and is now known to also grow in the Caribbean Sea near Jamaica as well as in certain areas of the Florida Keys.

Coral colonies rarely grow in isolation. In most cases vast numbers of colonies clump together into three basic types of coral formation. One type, the fringing reef, is an underwater coral platform, attached to a body of land and extending into the ocean for some distance. The barrier reef is another kind of formation that arises from the ocean at some distance from the shore and which, unlike the fringing reef, is not attached to the mainland. The highest parts of these reefs sometimes rise above the surface of the water and support a growth of trees, but I was told by Australian authorities that these tree-covered islands are merely tips of submerged mountains and are not formed by the coral animal, and that islands of the reef that have been formed by coral growth are called cays, and support no plant growth.

In America there seems to be a different interpretation of barrier reefs. An authority here says that many barrier-reef islands formed only of coral debris support a rather rich plant growth.

Two different American authorities have also disputed my use of the term *cays* as applied to the coral islands of the Great Barrier Reef of Australia. It is their belief that the term applies only to the reef islands of the West Indies. Yet, from the Australian books I read, and from my conversations with the inhabitants of Australia, I find they do indeed call them cays there.

The largest known barrier reef is the Great Barrier Reef of Australia. It extends 1,200 miles along the northeast coast of Queensland. It is fifty miles wide in places, and covers an area of 80,000 square miles.

In my companion book, I share with you my thrilling personal experiences as I explored the beauty, dangers, and surprises of Australia's stupendous Great Barrier Reef.

The third formation, called an atoll, is a coral island shaped like a ring. In the center of the atoll is a lagoon and often a channel connecting the lagoon with the outer ocean (Plate 7a). Such lagoons make safe harbors; however, many lagoons are filled with coral heads and shoal water, which prevents their use as harbors.

Sea Pansies

A truly remarkable sight awaits the explorer who has the rare good fortune to happen upon a bed of sea pansies (*Renilla reniformis*) under favorable conditions. It seems overhead weather conditions must be just right to coax them out of hiding when they are exposed by low tide, so they are seldom seen. They are a never-to-be-forgotten experience seen either by day or by night. Seen by night, they are enchantingly luminescent, and seen by day their lovely pink and lavender pansy faces so closely resemble the flowers for which they are so aptly named that one feels startled at finding them by the edge of the sea.

Though it is most difficult to believe, they are actually animals and not plants as one at first assumes. And what very odd animals they are! They belong to the same general group of simple beings as the jellyfish, anemone, and coral just discussed.

Their flat heart-shaped pansy faces are attached to short stems penetrating into the sand and, to avoid the drying rays of the sun, in clear weather they draw their beautiful faces just out of sight beneath the surface of the sand. They appear to be a single growth, but are actually a colony of many individuals that function in a most unusual way. The minute larvae are carried on the currents of the sea. When one makes a safe landing on the shore it ceases to be a single being and, through an extraordinary growth development, it becomes a group of many individuals. The various individuals, or polyps, are shaped like tiny tubes and are somehow bound together into a handsome flowerlike form.

Oddly, the tubes are not all alike. Some of them bear tentacles to capture their plankton food and to form reproductive cells. Other tubes lack the tentacles, and it is believed their function is to control the water intake. They appear to operate a kind of hydraulic system to change water pressure and thus control their movement.

How one longs to preserve their live fresh beauty for the collection cabinet. That is, of course, not possible. But the two I have preserved in my collection have held a portion of their color rather surprisingly well for ten years now.

4. Echinoderms: The Starfish Group

This chapter focuses on a group of ocean creatures familiar to all of us: sand dollars, sea urchins, sea cucumbers, and starfish. All belong to the phylum with the jawbreaker name Echinodermata, which comprises some 5,200 different species. This group of marine invertebrate (boneless) animals is distinguished, first, by a tough outer cover, or skeleton, composed of hard plates, or of hard formations called spicules, embedded in the body wall. Most, though not all, echinoderms are built on a radiating plan, often with five diverging parts as in certain starfish. They are all carnivores (meat eaters), with very simple digestive tracts, and have evolved over the centuries a very curious style of locomotion. The undersides of their bodies are lined with rows of tubular feet, tipped with tiny suction cups. Those tiny cups grip with such force that the animal can cling securely to a rock or cliff and resist even the strongest waves, or they can climb straight up the vertical side of a cliff or boulder. The suction cups are operated by a system of tubes through which seawater flows. Nature paved the way with her suction-cup invention, and man has now copied it to operate his hydraulic water pumps.

On the topside of most echinoderms' bodies, in the center, there is a dainty flowerlike design through which tiny lobed "feet" project. These feet, however, are for breathing and have nothing to do with locomotion or feeding.

Sand Dollar

The sand dollar (*Mellita testudinata*) (Figure 3) is typical of the echinoderm group, yet it is such an odd creature as to almost defy classification. Frequently called the Holy Ghost shell, it is quite common along the Florida coast. A fascinating legend is attached to the round, flat sand dollar that makes it one of the most widely known and popular of all shells. According to legend, the five slits on the edge represent the five wounds in the body of Christ. The dainty flowerlike design at the top is the Easter lily, with a five-pointed star in the center representing the Star of Bethlehem. On the back of the shell is the outline of a poinsettia, the Christmas flower; and when the shell is broken open, five perfect small replicas of a dove appear—the Doves of Peace.

There are several different species of sand dollars, each different from all the rest, and each found in its own special area and nowhere else. Just by looking at a sand dollar, you know at once, if you have collected along both coasts of America, whether it was found along the coast of Maine, Florida, Washington, or Mexico. Indelible ink is made from an exceptionally large and thin purple and yellow species found off the eastern coast of Africa.

Sand dollars are fascinating to hunt as they creep about in shallow water at low tide. With an odd twisting motion, they can quickly cover themselves completely with sand, so that often all you can see is a slight bulge in the sand. In life they are covered with brown fuzz. The snow white ones with which most people are familiar have been defuzzed and bleached.

Sea Cucumber

The sea cucumber, perhaps the least familiar of the echinoderms although there are as many as five hundred known species, belongs to the class Holothurioidea. It is one of the sea creatures that some collectors find beautiful, but to those of us with a revulsion toward anything resembling a snake or worm, its beauty must be enjoyed at a distance—a very great distance.

Some sea cucumbers can grow to an extreme length of seven feet, although the more prevalent species of these fat creatures reach only eighteen inches in length and three or four inches in diameter. They look exactly like long fat cucumbers or short fat snakes. They are often red or black and have a flowerlike ring of short branched tentacles encircling the mouth at the head end. When extended, these tentacles are beautiful even to the repulsed observer. They move about in slow fashion on a multitude of tubed feet and, at low tide, love to crawl back into crevices between and beneath rocks. Many times a collector is left shuddering in revolt when he turns a rock and realizes, too late, that he has actually touched one of these wormlike creatures.

When attacked, they have the uncanny ability to expel their own internal organs for the enemy to eat while they quietly crawl away and grow replacements.

The sea cucumber has few known enemies, but it has been recently discovered that the large species found along the Florida Keys and in the West Indies produces one of the most powerful of all known animal poisons, which is presumed to be a weapon of defense.

The sea cucumber acts as host to a tiny pearlfish (the fierasfer). The little fish takes refuge in the rectal cavity right alongside the sac of deadly poison to which, oddly, it is not immune.

Because of its prolific nature, great abundance, and feeding habits, the sea cucumber plays an important role in the distribution of sea-bottom deposits. With a crown of blunt tentacles operated by strong muscles, the sea cucumber scoops large quantities of bottom sediment into its mouth to extract the food particles from it, and then excretes the sediment. In so doing it loosens and aerates the soil of the sea bottom in a service similar to that performed on land by earthworms. Evidence from deep-sea dredgings has been interpreted by geologists as proof that great hordes of sea cucumbers plow up the sea bottom at very great depths.

The Chinese use great quantities of sea cucumbers to make a soup called trepang, which is one of their favorite foods. They are also widely used in Australia, under the name *bêche-de-mer*.

Sea Urchin

The sea urchin, which also belongs to the echinoderm group, constitutes a very large family of many different species. All are covered with spines of varying color, length, and sharpness, depending upon the species, and many are unbelievably beautiful.

The sea urchin is extremely prolific and plentiful in many areas of the world. It is frequently found crawling over rocks in shallow water at low tide. Those that prefer deep water often live in groups that sometimes almost completely carpet a large area of the sea floor. Some females lay up to twenty million eggs at one time. It seems to be God's plan that eggs and the very young supply a large part of the diet of sea life, so they are produced in tremendous numbers in the hope that enough will survive to perpetuate the species.

Timed in some mysterious way by its internal rhythmic nature clock, the urchin spawns each summer month at the time of full moon. All individuals of the species respond to this timing device simultaneously; thus the release all at one time of a prodigal number of reproductive cells is assured. Though the timing stimulus is not understood, it is clearly a provision of all-wise nature to ensure the perpetuation of the species.

On its underneath side, the sea urchin has a mouth armed with five teeth that can crush small mollusks, chew up vegetation, and scrape algae from rocks. This unusual mouth organ, composed of forty separate pieces, is called Aristotle's lantern. Sea urchins often swallow sand, grind it, sift out the food particles, and then expel the sand in a much finer form. It is believed that one species, *Echinometra lucunter*, called the rock-boring sea urchin, can actually chisel out a nest in rock and regrow replacement teeth as the old become worn. Often at low-tide line in a rocky area, the rock bank will be actually black as great numbers cling to the depressions they have each found or made for themselves in the rock wall. It is really a very beautiful sight as the water glistens on the black spines that protrude from their globular reddish bodies (Plate 7b).

Although it has no eyes, the urchin has an extraordinary awareness of approaching moving objects. Scientific study has revealed that its body is covered with a scattering of nerve receptors that are sensitive to changes in light intensity. If you approach the urchin from one side, all its spines swivel about on their little pivots to point directly at you in self-defense. If you then throw your shadow on its other side, it quickly swings about erecting its defense from that side.

The body of the full-grown black urchin grows to a diameter of four inches, with spines that often reach fifteen inches, and is poisonous to the touch. Contacting one of its slender hollow spines is as painful as a hornet sting and causes extensive swelling. For that I can personally vouch. The application of vinegar is a slight help, but it is largely a matter of just enduring for a day or so while you resolve to never again be so careless.

The long-spined black urchin (*Diadema antillarum*) is often found wandering about in clear view on sand flats at low tide. But more often it will be partially concealed as it nestles into a rock crevice with its spines all pointed out in defense.

The slate-pencil urchin (*Eucidaris tribuloides*) has beautifully colored short fat spines that seem an exact replica of the slate pencil for which it was named. It is a sluggish, slow-moving creature, and seems dull and unaware of danger. Most urchin species offer resistance and cling to an object with their tube feet when you pick them up, but the slate pencil offers no resistance at all. In the Florida Keys area, it is usually seen one at a time and is believed to be of a solitary nature. But one of the outstandingly beautiful underwater memories that I treasure is a colony of at least a hundred or more on the ocean floor about a mile offshore from one of the islands of the Hawaiian group.

The slate-pencil urchin family is believed to be the only modern urchin that also existed in Paleozoic times. The recent members of the group show almost no change from their ancestors that lived hundreds of millions of years ago.

While searching the high-tide litter line, one often finds the lovely little empty home of the heart urchin (*Moira atropos*). None are ever seen alive on the surface so they inspire curiosity. This odd little creature builds a subterranean chamber for itself. Connecting its chamber with the floor of the sea, it keeps channels opened and lined with sticky mucus. Through these channels, it reaches up to retrieve plankton and other food particles lodged among the grains of sand.

This section cannot rightfully be closed without special mention of the lovely green urchin (*Strongylocentrotus droebachiensis*). It thrives best in rocky, cold northern regions, and is found especially in great abundance along the coast of Maine. The green urchin seems to have an affinity for rock surfaces covered with a crust of rose-colored coralline algae. They cling by the thousands to rock surfaces along the cold northern shorelines.

With their hundreds of little tube suction-cup feet, they cling with such tenacity that they cannot be dislodged by the strongest of waves. Apparently to protect themselves from their mortal enemies, the sea gulls, they choose such a low beach elevation that they are very rarely exposed by low tide.

The eggs of the sea urchin are considered by some to be especially good eating. In Barbados the commercially canned eggs are served in hotels. As a means of increasing my education, I usually make it a practice to at least taste every edible thing from the beaches. I was far south on the western coast of Mexico, just starting to prepare my first urchin egg meal, when a local family waded ashore with a basket full. Acting half starved, they eagerly and immediately ate the eggs raw—no washing, no cooking, nothing. Just watching, the "butterflies" started fluttering in me until even now I am unable to taste one. So I cannot personally vouch for their edibility.

Unfortunately, the beauty of the sea urchin cannot be preserved for the display cabinet. The pretty colors fade, and the spines retract from the upright position and, in many cases, fall off entirely. Every collector has his pet method, but I have never yet seen one perfectly preserved. The method I feel is most successful is described on page 260 in Chapter 17.

Starfish

The starfish, just as the sea urchin, sand dollar, and sea cucumber, is an echinoderm with its skeleton composed of limy plates. It has no resemblance to a fish, and is more correctly called the sea star (Plate 7c).

Almost all tropical seas abound with many different species, and some species grow as far north as Greenland. Over seventy species inhabit the Puget Sound area alone. Most have radiating arms, five being the most usual number, although a few species have many more. Some are very beautifully colored with gaudy shades of red, blue, orange, purple, and so on, while other species are drab and dull. In size they range from extremely tiny to a forty-inch width. Most have a tough, leathery skin covering the limy-plate skeleton. When the skin is removed, the skeleton is surprisingly pretty, with the series of limy plates beautifully joined together.

On the underneath surface, the mouth is in the center of the central disk, with the digestive system extending into each arm. At the end of each arm is one or more flexible sensory organs with which it detects vibrations and chemical substances in the water and, except for the deep-sea species, all starfish have a small cluster of simple eyes at the tip of each arm. Numerous tubular feet project and retract through the many perforations in the limy plates along each arm. At the end of each tube is a tiny suction cup to aid the animals in walking up to two or three inches per minute. The extremely large species have as many as ten thousand suction cups with a combined hundred-pound gripping force. The tube feet connect with a system of internal tubes through which sea water flows and operates them by a highly perfected hydraulic suction pump.

The starfish lives chiefly on clams, that is, bivalve mollusks, and it kills and eats its prey in a most unusual way. Although starfish move slowly, mollusks move even more slowly, so catching them is no problem. The starfish mounts upon the clam in a humped-up position, attaches its tubed feet to the two shells of the bivalve, and begins to pull. The clam reacts by closing its valves tightly. Anyone who has ever tried, bare-handed, to open a live clam will wonder how it can be done by a starfish that is often no larger than the clam it attacks. The secret is that the starfish uses the numerous tube feet in relays. It is, therefore, able to outlast the clam and maintain its grip until, from exhaustion and a need to breathe, the clam opens its shell just a little. The stomach of the starfish then turns inside out as it comes out through the mouth and enters the partially opened shell. There it secretes its digestive fluids, which quickly break down the clam's meat so it can be drawn back into the starfish for the completion of digestion.

For my collection, I was fortunate in being able to capture and preserve one in the act of eating a mussel.

The starfish, with its huge appetite, is extremely destructive to the oyster, clam, and mussel industries. A small three-inch starfish eats three or four oysters daily. Often a large school will move in and destroy entire crops before the seafood farmers even know they are there. The farmer must continuously fight them by dragging tangle mops over the planted beds. They are equally destructive to all kinds of shell life in the sea. Mollusks recognize their danger by instinct. The minute a starfish appears on the scene, shellfish start scurrying for cover. The seafood farmer is overwhelmed by the never-ending multitude of starfish just as the corn farmer is overwhelmed by weeds, and each is mystified as to how they can prevail in such numbers. But with a closer look at nature's "starfish factory," at least that part of the mystery begins to clear up.

A female starfish is believed to produce about ten million eggs a year, and all species of the family have the peculiar ability to regrow a new arm, if one is lost by injury. Some species can even deliberately cast off an arm that may be in the grip of a predator. I have one in my collection that lost all but one arm and was successfully growing four replacements. Some of the star species can be cut squarely through the middle of their bodies and each half will regenerate to become a whole shell again. And even more remarkable, single detached arms have been known to regenerate a whole new body. In times past, before the regenerating ability of the starfish was discovered, oyster farmers made the mistake of cutting the attacking pests in half to kill them, thereby unknowingly compounding their troubles.

Every creature, however, has its redeeming feature, and for the collector of miniature mollusks the starfish is a most efficient helper. At the base of the arms are small pockets that sometimes bulge with tiny miniature shells. I have known of as many as twenty-four being removed at one time before being consumed by the digestive fluids of the starfish. These small pockets appear to be some kind of food storage unit whereby the starfish can accumulate periodic feasts.

In my opinion, the most beautiful and unusual of all the starfish is the basket star, though it is seldom seen (Plate 8). The basket star has a very small central disk and its twenty-four-inch arms are finely branched and coiled so that when they are spread out they look for all the world like beautifully designed fine tatted doilies. During the night, they spread out their complicated array of tentacles to catch plankton for food. By day, they curl into a ball and retire into a crevice, or else entangle themselves in the branches of the sea-feather plant where they become invisible. In all my years of looking, I have seen only one spread out and it was simply fantastic. I have one in my collection, but it is curled in the "daytime" ball so it does not tell the full story.

Starfish in general are comparatively easy to find but are difficult to preserve for a display cabinet. They are found in tide pools and crawling all about on rocks at low tide. But, as far as I know, no way has ever been devised to cure them and fully preserve their original brilliant colors.

Crown of Thorns

For a long time, areas of dead coral heads had appeared on Great Barrier Reef of Australia, and no one understood why. It was eventually discovered that the large fifteen-inch spiny starfish (*Acanthaster planci* Linnaeus), called the crown of thorns, was responsible. It was witnessed in its act of destruction, which removed all doubts. It was seen with its stomach inverted, digesting the living coral polyps. When its meal was finished, it moved on, leaving one more dead coral head in its wake. Great concern was felt over the future of the Great Barrier Reef. The Arlington Reef, five miles long and twelve miles wide, was reduced to rubble in just a few months' time. Later the popular tourist area of Green Island, off the coast of North Queensland, was severely attacked.

When the deadly crown of thorns first appeared in the popular Green Island tourist area about 1960, they were such a novelty that the glass-bottom boats made regular trips to show them to tourists. It was not until 1965 that they had multiplied to plague proportions. The glass-bottom boats were then forced to travel over several miles of destruction in search of undamaged coral for the tourists to enjoy. It was estimated that 90 percent of that beautiful reef area was destroyed in a matter of months.

Divers were then employed in the hope of rescuing the Green Island reef for their tourist operations. In three years time it was estimated that more than 44,000 starfish were lifted from the water and buried on shore in the sand. Their needle-sharp spines can penetrate leather gloves so they must be handled with great care.

Compounding the tragedy was the follow-up disappearance of the unusual and exotic reef fish. Apparently the living coral is essential to their livelihood, for as soon as the coral is killed, the fish disappear.

The area of invasion, in 1968, was found to extend from the Capricorn Islands off Gladstone, Queensland, as far north as Cooktown, with the more northern areas suffering the greatest damage. By 1969 it was estimated that the prolific starfish had killed a hundred square miles of the living reef.

By the fall of 1969 it was discovered

that the phenomenal population explosion had gone far beyond the Australia area. Divers found a heavy concentration of the crown of thorns off the Hawaiian Island of Molokai. Dr. Albert H. Banner, a University of Hawaii zoologist, reported that they had upset the balance of nature throughout hundreds of the South Pacific islands. Ninety percent of the northwest reefs of Guam had crumbled, and the reefs of Hawaii were under severe threat; a tragedy which, if not prevented, would have left her beautiful beaches at the mercy of the pounding surf.

Scientists from several nations set up twenty island laboratories from Hawaii to Australia to determine the cause of this rapid increase in the starfish population. Four theories were advanced. First, that the plague resulted from the unexplained mass mortality of a certain fish species that had formerly preyed heavily upon the starfish. Second, that the gradual warming of the surrounding waters might have induced its more prolific growth. Third, that ocean pollution has killed the coral organisms that once fed on starfish larvae, and fourth, that excessive commercial harvesting of its natural enemy, the triton trumpet shell, might be responsible (see page 172).

All starfish have been created with an amazing ability to perpetuate the species, so the task of trying to control the destructive overpopulation of the crown of thorns seemed monumental indeed. Then came a second great mystery.

While scientists were still scratching their collective heads trying to figure out the what, why, and how, all-wise nature stepped in and seemed to quietly solve the unsolvable problem. By 1972, according to television and newspaper reports, the starfish seemed to disappear, while the mystified collective head-scratching still continued as the experts then tried in vain to solve the what, why, and how of their apparent departure. But alas! It was too soon to relax and the head-scratching still goes on. By 1973 there was a massive invasion of another area. The alarmed inhabitants mobilized for action, used bulldozers to dig big holes at intervals along their threatened beaches, and captured and buried some 47,000 of the destructive invaders. Time will tell.

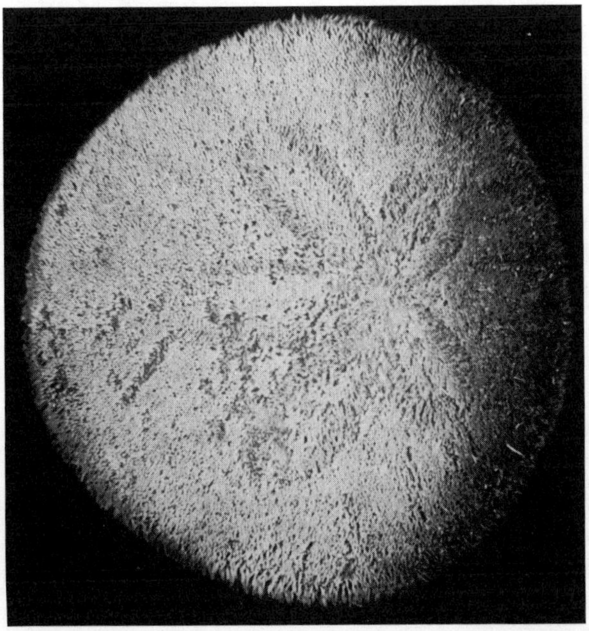

Figure 3. A Mexican sand dollar in its natural state.

5. Crabs and Other Crustaceans

The succulent crabs, lobster, and shrimp that often grace our dinner tables belong to the class Crustacea, a large group of more than sixteen thousand known species. Crustaceans, like mollusks, are gilled invertebrates whose soft boneless bodies are protected by a hard limy shell covering. Unlike the mollusk, however, whose body is all of a piece, the central part of a crustacean's body is segmented, with jointed legs, or appendages, attached to the torso.

Crustaceans play an important role in the ecology of the oceans; they constitute the chief food of many valuable food fish, and are the natural scavengers of the sea. Since their flesh is very tasty, and in great demand as human food, their commercial value is also great, and has become greater in recent years with the advent of the deep freeze. Formerly, most merchants demanded live delivery, since crustaceans spoil very rapidly after death; this confined their use largely to coastal areas. But now inland peoples as well can taste these culinary delights.

Crustaceans do, however, have their nuisance value too. They are one of the most persistent enemies with which shellfish farmers must struggle.

Among the many types of crustaceans, it is crabs that are most frequently encountered on the sands and in the shallow water of ocean beaches. So in this chapter we will concentrate largely on crabs, and on the Maine lobster.

Crabs

There are more than a thousand known species of crabs, and they vary in size from microscopic to the giant *Macrocheira kampferi,* which sometimes measures eighteen feet across the outstretched legs. There is just as wide a variation in color and modes of living. In all species, however, the body is broader than long, the mouth is very strong-jawed, and the stomach is lined with internal projections, or teeth, for grinding. Potential predators are discouraged by the crab's hard shell. Indeed, some crab species have evolved the clever defense technique of leaving an arm, or claw, in the grip of an attacker, while going away to grow new ones.

Periodically crabs molt, throwing off their entire limy covering. They are then, for a time, soft and helpless and search out some secluded nook to await the growth of a new, and larger, replacement shell. In this state they are the soft-shelled crabs of the market. The tender meat of many species is favored by people all over the world, which has made crab fishing and canning important world industries.

Blue Crab

The blue crab is the best known and most valuable of the swimming crabs (Plate 9a). Although especially popular in the Chesapeake Bay area, it is highly prized as food all along the Atlantic and Gulf coasts where it thrives. Unlike most crab species, its last pair of legs is flattened and adapted for swimming.

The female blue crab lays her eggs in summer and attaches them to the outside of her own body. In about two weeks they hatch as larvae, in a form very unlike the adult. They alternate growth and molting and in time assume adult form. The adults are caught and marketed during the soft stage between molts when they are most palatable.

Fiddler Crab

The common and plentiful little fiddler crabs (*Uca minax*) are most amusing to watch on the beach (Plate 9b). They travel in great numbers and frequently run with amazing speed in a peculiar sidewise motion. Catching specimens is real sport, for you must exert skill to outmatch theirs.

The fiddler crab differs from most crabs in that the male has a huge claw on one side and a small one on the other, while the female has only two small claws.

These crabs prefer the higher areas of sandy beaches and salt marshes and feed on organic matter in the sand. A colony of the small crabs digs burrows up to three feet long. Water covers most of the burrows at every high tide. As the water rises, the crabs plug the entrance and wait for the return of the next low tide to release them from their prison.

The fiddler crab figured in a very interesting set of biological clock experiments, with startling results. Researchers set out to prove that this crab, which is dark during the day and blanches to a pale color at night, alternated colors in response to the alternating day/night change in light intensity. But the exact opposite turned out to be true: Even when the crabs were subjected to a constant light environment with no apparent day or night, the rhythmic color changes never ceased, triggered by the crabs' internal time clocks.

Great Barrier Reef Crabs and Shrimp

The creatures that inhabit the warm waters of the Great Barrier Reef of Australia are noted for the extra brilliance of their color, and Barrier Reef crabs, lobster, and shrimp are no exception. They are beautifully hued and add much to the color symphony of the reef. Through the night they are very active in their search for food. By day they are much more difficult to see since most species hide in some crevice to await darkness.

The Great Barrier Reef, along with coral reefs throughout the world, is also the home of a small shrimp with a most unusual feeding habit. It is equipped with a long antenna, which it flips about as it sways back and forth to attract certain fish species. Then, with full cooperation from the chosen fish (usually barracuda and moray eels), the tiny shrimp proceeds to clean the teeth of the fish and feast on the food lodged in their teeth.

Other species of shrimp and certain species of colorful fish on reefs pick small parasites from the fins, scales, and gills of fish, congregating into cleaning stations to which the troubled fish come to receive this service (Plate 9c).

Hermit Crab

Of all God's beach creations, certainly none is more utterly fantastic than the hermit crab (*Pagurus*). It would seem the great Creator was somehow interrupted in His task and abandoned the poor thing when only half finished. The front part of the crab is all nicely completed with claws, eyes, feelers, and every equipment needed for living. But the back half of its body is quite unfinished, naked, and vulnerable.

It would seem that eons ago when God, for some unknown reason, walked away with His task unfinished, the poor crab must have seen an empty seashell lying nearby and hastily backed into it to cover its embarrassment. So nicely did the shell serve its purpose that it has continued backing into it on down through the centuries (Plate 10a).

Everywhere in the world, wherever there are empty shells on the beaches and in shallow coastal waters, they are inhabited by the half-formed hermit crabs—all sizes, from the very tiniest of shells to large ones. As soon as the hermit crab feels the pinch of its chosen shell on its growing body, it chooses another of a slightly larger size. Occasionally it sets its fancy on a shell with the living mollusk still inside. Undaunted, it soon tears the mollusk to shreds, pulls it from the shell, and claims the home for its own. Once inside the shell, it closes the opening with an odd pincer arrangement created for the purpose. With that claw drawn tightly in place, its door is closed as securely as it was when the original mollusk drew down its operculum.

The hermit crab is far more active than the original occupant. Beginning shellers often think, when they see a shell being dragged around by a hermit crab, that they have found a live mollusk. But with a little experience, the difference in speed is a dead giveaway.

The hermit crab, a beach scavenger, serves an important function by feeding on decaying matter that would otherwise foul the beach. Though its feeding habits are repulsive, it is otherwise a very tidy creature. Using its claws for scraping and rubbing, it constantly grooms all parts of its exposed fore body, refusing to tolerate any encrustment of marine growth. The small crabs are drab and colorless, but there is quite a huge one in my collection brilliantly trimmed with gaudy colors. It is really very lovely. By some kind of evolutionary process, a few hermit crabs have developed an air-breathing capacity and can make land excursions, even climbing trees. The mangroves along the coast are their favorite haunt. It gives the inexperienced collector quite a start the first time he sees an apparently live saltwater shell climbing a tree.

Horseshoe Crab

The horseshoe crab is not actually a crustacean at all—it is really one of the spider and scorpion group—but since it is universally referred to as a crab, we will include it in this chapter. Through the study of fossils, it has been determined that this animal has survived, unchanged, since the earliest Cambrian age. Only five known species have survived to the present time: our eastern North American species (*Limulus polyphemus*), found from Maine to Yucatán, and four other species along the eastern coast of Asia.

Our common American horseshoe crab is frequently called the pan crab because its shell was used as a pan by the early settlers. It is truly one of God's beach oddities, and a rather frightening one at the first encounter. As it glides toward you with that inverted pan-shaped shell covering the entire animal, it looks like nothing so much as a small armored tank approaching with no visible means of locomotion. As you gaze in wonder, you are soon overcome with the startled feeling that some weird creature is underneath the tank peering at you through the two little isinglass windows at the front (Plate 10b).

The horseshoe crab ranges from shoreline to a forty-foot depth, and is believed to live in the same beach area throughout its life, although some have been known to travel as much as twenty miles.

In spring and summer, these crabs move to gently sloping beaches in shallow bays to spawn. The small males move shoreward first and wait in a stag line for the much larger females. With a hook and groove system, the small male hooks two especially built claws to the rear edge of the female's shell and hitches a ride to the egg-laying beach. Arriving at the beach, the female scoops out several depressions. Into each hole she deposits several hundred eggs, and the male releases his sperm on the eggs as she drags him over the nests.

The eggs, about $1/12$ inch in diameter, are transparent and greenish blue in color. In her several nests the female lays as many as ten thousand eggs, about one-half pint, and they are quite edible for humans, as well as being highly relished by sea gulls and other seabirds.

The eggs in each nest are covered with very little sand and require from three to seven weeks to hatch. The larva then, subsisting on its egg yolks, crawls, swims, and floats away from the beach into shallow water. In a short time it molts, emerging with all the physical characteristics of the adult. It molts several times the first year, and then only once each year until it stops growing at about ten years of age. The last shell is almost black and well protected with thorned edges. The adults reach about a two-foot length, including the tail spike. Many people fear the tail, but it is actually not a defense weapon, and oddly, if it is broken off, a new one grows to replace it.

As an adult, this odd creature is adapted for living in soft sand and mud at the bottom of the sea. Endowed with a special ability, the horseshoe crab has for 300 million years been plowing furrows at the bottom of the sea; it plowed for many centuries at that great depth before man finally copied and applied its plowing principle to the soil on the surface of continents. It uses five pairs of strong ambulatory legs to push its body forward, plowing a furrow through the mud as it searches for worms and other buried small animals. Unusual toothed projections jut out from other appendages. As the crab moves forward, it uses the projections to shred the worms and other prey before stuffing them into its mouth.

This crab will, however, sometimes choose to swim instead of walk, flipping over on its back and using its shell as a boat, and its legs as oars. But when the water turns cool, it burrows for protection into the sand and mud.

The horseshoe is not an edible crab species, but it is of considerable commercial importance. It is not only valuable eel and lobster bait, but huge quantities are used in the preparation of fertilizers and feeds for poultry and pigs. But perhaps its greatest contribution to humanity is an unusual invention nature built into its eyes. Its eyes are highly developed with very special features much too complicated to explain here. The General Electric laboratories have now used the principle of one of those features in perfecting a new system to sharpen television images. And it is hoped that with better understanding of another of its special features, its eye can be used to perfect a compass that will function at the North Pole, where our present magnetic compasses fail to function.

Land Crab (Florida)

The large and interesting land crab (*Cardisoma guanhumi*), so widespread in Florida, is actually not a land crab at all, although it spends most of its life as much as five miles from the sea. The mysteries of nature are truly beyond understanding. This odd creature must have salt water to live, so it packs up its own private salt factory and carries it inland to make its own supply. Finding just the right spot, it proceeds to build its home by burrowing down to the underground water table, often four or five feet down. If it happens to be a saltwater table, its home building task is completed. But if it is freshwater, it operates its little salt factory, releasing from its own body exactly the amount of salt needed to increase the salinity of the water to the right concentration.

The pool, which it has acquired at such effort, is actually not its home for it is a so-called land crab. It is merely its bathing pool where it goes from time to time to bathe its gills. It will die within a few days if denied access to its bathing pool.

These odd crabs also have the peculiar ability to survive for months without food. Yet, when food is available, they can eat one-fourth their weight in one day. They are primarily vegetarians and cause much damage to cultivated crops, especially along the southeast Florida coast where they are very prevalent. They are especially destructive to Florida's tomato crops, and related species of crabs destroy many acres of rice in the Far East. As many as 7,500 crabs have been counted at work on one acre of land.

Other favorite foods are the leaves and fruit from the mangrove, buttonwood, and strangler fig trees. Both their vision and hearing are keen enough to detect the falling of a leaf. Some colonies live along canal banks feeding on the algae growing on the water and on dead fish thrown away by fishermen.

In the Indo-Pacific and Central America land crabs are widely used as food. But, until recently, because of an objectionable odor of iodine, not many were eaten in this country. It has now been discovered that a delicious odor and flavor can be produced by feeding them lettuce, bread, and watermelon rinds for a time before killing. With that discovery, they are now gaining favor as food in the United States. Their egg masses are especially prized, since they taste very much like shad roe.

Once each year, at spawning time, these crabs make a mass migration to the sea. They move in fabulous numbers and startle many a motorist as they cross the highways of southeastern Florida. Hundreds are killed by the cars and many car tires are punctured in the process. The migration lasts from June to December, reaching its peak in September and October. The major movement occurs each month during the three days preceding full moon.

Mating takes place in the burrows before migration begins. Then both sexes head for the sea, with the female carrying thousands of fertilized eggs attached to abdominal appendages. She releases her young in the shallow water at the edge of the sea with a fanning movement of the abdomen. After hatching, the tiny, free-swimming crabs remain in the sea for a time, continuing their growth. Many fall prey to larger animals, and the small percentage that survives later crawls ashore and moves inland with their little salt factories to carry on life as their ancestors have done for thousands of years.

Mole Crab

These odd little surf fishermen (*Emerita talpoida*) are truly one of nature's fascinations to watch. They live in colonies of literally thousands and fortunate, indeed, is the observer who is privileged to watch them in action. You can be wading near the water's edge feeling the beach is quite destitute of all life. Then, as a retreating wave goes back to sea, you suddenly realize that scattered all about you in fabulous numbers are tiny long-whiskered, beady-eyed faces peering up through the sand.

The small molelike creatures, with their flattened appendages, are well named. Endowed with many sensory bristles, they have a keen sense of touch, but it is believed their small eyes are practically useless. They are equipped with long, curling, feathery antennae, which they use most efficiently to go fishing for their plankton food.

To appreciate the miracle of the mole crab, one must first grasp the seemingly impossible problem that it must face—and solve—if it is to survive. The little shore creatures no longer than your thumb must do their fishing in the incoming rolling, ever-moving surf, yet they must not permit themselves to be carried out to sea by retreating waves.

Their solution to this problem is not only ingenious, it provides the observer with an entrancing sight to behold. With their bodies buried vertically in the sand, and only the tiny faces peering up, they stand, as it were, at attention. Then suddenly, at the exact same moment, just as if all were responding to a command from a leader, every little face sends up its long feathered antenna fishnet, as a small wave of water rolls gently over them. They then draw the plankton-laden antennae through appendages surrounding the mouth to scrape off and eat the captured food.

It is an amazing sight, but a little later something happens that is even more extraordinary. As the tide starts to come in and water depths change, they must retreat up the slope of the beach. They wait until an incoming wave is rolling forcefully over the bed. Then, again in unison as if by command, they all pop out of the sand until the sand seems to be boiling and permit themselves to be washed higher up the beach slope. During the short pause needed for the water to reverse its course for the return trip to the sea, with amazing speed and magical ease they dig backward into the sand with a whirling motion of their tail appendages. And once more they are ready to cast their fishing nets with the water at correct depth to suit their fancy. To maintain that correct depth at all times, they must make several mass upward movements during each incoming tide, and then again a like number of downward movements during each outgoing tide.

Since these small creatures must live at the very edge of the surf, they have a double set of enemies. They are an important source of food for larger animals of both land and sea, thus they help supply a vital link in the chain of events that converts the basic plankton of the sea into food for man.

Those that survive live about eighteen months. Assuming that no predator cuts their lives short, they spawn and, for several months, the mother carries a mass of orange-colored eggs firmly attached to her body. As hatching time nears, the mother stops feeding and remains at the low-tide line, thus avoiding all risk of her offspring being stranded in the dry sands of the upper beach area.

After hatching, the female dies (the male dies soon after maturity), the larva enters the water currents and, through its early stages of development, is but another of the many creatures in the plankton of the sea. The larva is quite long-lived and often is carried many miles from the beach of its origin before it settles on a beach to continue life in the manner of its ancestors.

Sponge Crab

The sponge crab (*Dromidia*) is an extremely shy member of the crab family. So timid is it that it wears a disguise at all times.

Two of its legs are especially designed for clasping. It cuts a piece of sponge exactly the shape of its top shell, and then, with its clasping claws, holds it firmly over its back when it ventures forth from its nest.

Its home is the original sponge from which it cut its covering. So carefully does it cut the sponge covering that when it retires from its wandering, it crawls into the cavity in the mother sponge, pulls the covering into place like a plug, and completely disappears from view.

Crab Oddities

The female of one tiny crab species (*Hapalocarcinus*) settles on living coral soon after hatching and is soon entombed for life within the hard growing coral. She is fertilized by the free-swimming male shortly before her complete imprisonment, and the babies escape through small pores in the coral wall. Seawater flowing in through the porous walls carries plankton food for her survival.

Another odd little crab, Lybia, with a rather soft shell and very feeble legs, lives in the Indian and Pacific oceans, with apparently no means of defense against its many enemies. Sensing its own helplessness, it carries a poison sea anemone in each claw, and nothing can attack it without receiving a painful sting.

There are many other interesting species of saltwater crabs that add zest to life for the beach explorer. For example:

Mud crabs, which hide by changing color to match their surroundings.

Green crabs, which like to swim about in brackish water.

Calico, or lady crabs, with speckled, almost square, shells, are sometimes caught for food.

The tiny Portunus crab, sometimes found on seaweed.

Rock crabs, which are of considerable food value, like both rocky and sandy beaches. Some species grow a foot across.

Sand, or ghost crabs, which move so fast on dry sand they seem to disappear right before your eyes.

Spider crabs, with their long gangling legs and their backs covered with algae, barnacles, and sea anemones.

Alaskan king crabs, which grow to twelve-foot lengths, are one of Alaska's big seafood industries because of the popular demand for their delicious flesh.

Kelp crabs, a very small species of the spider group found in kelp beds and tide pools.

Maine Lobster

The very delicious and popular Maine lobster (*Homarus americanus*) is facing an unusual threat of extinction. It is peculiarly adapted to the exact range of water temperature that, for years, has prevailed along the coast of Maine. In 1967 a sudden several-degree drop in water temperature began to threaten the lobster industry. To preserve the lobsters and other endangered shellfish, re-

Figure 4. The bigger lobster, an unusually large two-foot-long specimen, was captured off the coast of New England. The smaller one, weighing only 1 pound 2 ounces, is about one foot long. *Rex Gary Schmidt, courtesy U.S. Fish and Wildlife Service*

searchers are now planning a fantastic experiment to be carried out in Casco Bay on Cousins Island, Maine. While protecting the lobsters with rocks and cement blocks, scientists are attempting to recruit a high enough concentration of lobsters in the one area to interest the federal government in financing the remainder of the experiment. And what is the experiment? You would never guess it. The researchers hope to warm the waters of shoreline coves with heat discharged from power plant cooling systems. They even hope to somehow overcome a law of nature and keep the heat in the bottom layer of water where the lobsters can benefit from it.

Even before the water temperature problem arose, the lobster was in serious danger from overfishing. At the time of the early white settlers, our North Atlantic coast was literally crawling with lobsters. Some had grown to a huge size. One thirty-five-pound lobster, with a record twenty-four-pound claw, was caught in Massachusetts. (The male has one claw that grows out of all proportion to the remainder of his body.)

From the very beginning, due to the popularity of the delicious meat, lobsters were fished far too intensely. The yearly catch built up to 100 million pounds by 1885, and by 1918 lobsters were so scarce it was feared they would become completely extinct. The government stepped in with conservation laws and the supply improved, but the era of the large size ended. Ninety percent of all lobsters reaching legal size are caught each year by the eight thousand United States fishermen and the twenty-five thousand Canadians operating four million lobster traps. Since the advent of freezing, they are now popular in inland areas and there is no limit to the demand.

Of all nature's creatures, the Maine lobster is one of the most unusual (Figure 4). It walks forward on eight slender legs, but when it is frightened it flips its tail and makes a quick backward dash. It has six mouth parts and chews sideways. Its kidneys are in its forehead. It listens and smells with its legs. It tastes with its feet. Its teeth are in its stomach. Its brain is in two parts—one above and one below the throat. Its nervous system is along its belly instead of its back. It is nearsighted. For seemingly no reason it sometimes dismembers itself. It gorges itself until it splits its shell, then crawls out of it and grows a new one. It presents the lady of its choice with sperm cells which she carries for months in a sac alongside her latter two pairs of walking legs. She then curls her tail to form a pocket, lays her eggs in the pocket, cements them fast, releases the sperm on them, and carries them about a year before they hatch.

Lobsters, like many forms of sea life, hatch in great numbers. Seventy-five thousand is not unusual for one brood. Ninety-nine percent are eaten by fish and other lobsters as soon as they hatch. About one in a million has a chance of reaching adulthood. The few that survive sink to the ocean floor in three or four weeks where they have a better chance to hide from their multitude of enemies. If given an opportunity, the lobster has the capacity to live for fifty years and grow to the huge size described. But it seems that its meat is simply too tasty for it to be granted that opportunity.

Barnacle

The barnacle, assumed by most people to be a mollusk, is actually a highly modified crustacean of the order Cirripedia. (The term *Cirripedia* means feathery, or hairy, legs.) It is quite different from crabs and other crustaceans except for its jointed appendages, which seem to be the determinative factor in classifying it as a crustacean. The barnacle remains permanently attached for life while all other crustaceans are ambulatory.

There are 720 known species distributed worldwide. It is estimated that a single mile of shoreline can support a billion or more barnacles. As many as 150 of the smaller species may live on one square inch of rock surface.

The barnacle is hermaphroditic (both male and female organs) and is extremely prolific—since all barnacles produce eggs. These thousands of eggs are held within the parent shell for several weeks until hatching time, when the young free-swimming larva is discharged into the water.

At first it bears no resemblance to the parent. During the first month of life, it passes through six growth stages as it swims about eating the microscopic plankton of the sea. Then, guided by some uncanny instinct, each individual larva chooses exactly the right place to settle where conditions are best adapted for the survival of that particular species—some in the light, some in shadow, some high on the bank, some low.

Settled on some rock, ship, wharf piling, or floating object, the barnacle then passes through a complete metamorphosis. It cements itself fast for life, sheds its eyes, and grows the form characteristic of the adult of the species. A hollow, crustlike form of overlapping shelly plates, very much like a tiny bird's nest, builds around the animal. When retracted into this shell it very closely resembles a young bird's broad beak settled down in the tiny, hard, calcareous, limelike nest. When it feeds, that broad beak, consisting of four valves, opens up like double doors and the many-jointed hairy appendages protrude above the edge of the nest and sweep through the water like a casting net, entrapping small animals and organic fragments. The ever-hungry animal eats almost any living thing, either animal or plant, from a quarter-inch size down to microscopic.

One soon learns not to sit on a pier with one's bare legs hanging down in direct contact with barnacles. Those appendages come out to partake of your legs and soon have your skin red and itching. I do not know whether the irritation is caused by biting or by secretion.

Whales are caused great distress by the large and very beautifully designed whale barnacle (Plate 10c). It is not a parasite, but merely attaches itself to the whale's skin. I had my first encounter with whales and their barnacles while out on a boat with an old sea captain who was devoting his life to the study and collecting of marine specimens for scientific research. We saw a whale thrashing about near our boat, jumping from the water and throwing its body about. The captain explained it was being tortured by barnacles and trying to break them loose. He said that in their distress they often lashed against exposed rocks or steep banks, tearing barnacles away, skin and all. He gave me some of the whale barnacles for my collection, which, in my opinion, are the most beautiful of all the barnacles.

As with many other things, there is no harmony of agreement on the captain's belief. I have also been told the whale was jumping because it was the mating season. To me, both explanations sound possible and logical.

The lowly barnacle can be listed both as an asset and as a liability. It is an important item of food for worms, crabs, fish, sea slugs, and so on, and the larger species are an important source of human food in China, Chile, and many other areas. But it is a great pest to the shipping and dock industries since great sums of money must be spent fighting them.

It is believed that a barnacle lives from one to seven years, depending upon the species. When a barnacle dies its shell remains attached to the rocks and is used as shelter by many small creatures of the shore. The adhesive excretion with which the barnacle attaches itself to foreign objects is so superb that it has excited the interest of scientists who are struggling to duplicate it. It is hoped that at some future time dentists may be able to use the synthetic glue to permanently fix fillings in teeth. According to the findings of the National Geographic Society, the glue produced by barnacles has twice the strength of any commercial glue now on the market. The natural adhesive hardens in salt water and withstands a pulling force of more than seven thousand pounds.

The chief enemy of the barnacle is the common dog whelk, or *Nucella*. The brightly colored carnivorous dog whelk prefers barnacles to all other food, perhaps because it need not first bore a hole through the shell; it just envelops the cone within its foot and forces the valves open. The whelk then secretes a substance called purpurin, which seems to have a narcotic effect and eases the task still further.

According to an old Irish myth, in northern Europe and Greenland there is a small seabird called the barnacle goose, which very closely resembles certain species of barnacle. From the eleventh to the seventeenth centuries, when its remote Arctic nesting places were unknown, it was believed the goose had developed from the barnacle. But not knowing whether a bird so born should be considered fish or fowl, great controversy arose among theologians as to whether the animal could be eaten on Fridays and other fast days. Anyone interested in the complicated origin of this myth should read *Barnacles in Nature and in Myth* by Heron-Allen.

6. Fish and Fish Curiosities

In this book of unusual nature facts, it might seem that fish are hardly worthy of discussion, since everyone knows what a fish is like. Even people far from the ocean have seen fish swim about gracefully in their watery home. However, few people have ever stopped to wonder how fish "work:" how they swim, how they regulate the depth at which they swim, how they keep from becoming exhausted when they are on the move for such long periods of time. How do deepwater fish, who swim in total darkness, know which side is up? And most important, how have they managed to survive, surrounded on all sides by mortal enemies?

Along with mammals, birds, and reptiles, fish belong to the subphylum Vertebrata: animals with a spinal column and an internal bony skeleton. Fish, however, differ from most land vertebrates in that they are cold-blooded, they breathe with gills that extract oxygen from water, and they adjust their weight in water by means of gas swim bladders. So extremely hostile is their watery environment, that the fish's ability to have survived over the millennia is beyond comprehension. The miracle of a fish could not possibly have been achieved by the inventive genius of man. But, with the guiding hand of nature, it has managed to solve all its problems and maintain itself down through the centuries.

Before taking a close-up look at a few special fish oddities, let us first view more carefully that miraculous equipment with which all fish are endowed. For ease of swimming, a fish's body tissues, bones, air spaces, and so on are subtly constructed so that its body weight is the same as the weight of the water its body displaces. Thus, with its specific gravity equal to that of the water, it remains suspended in it with no effort. And, so that it might move forward through the water with the absolute minimum of friction or resistance, its shape is pointed at the end to break and slide through the parted water with the greatest of ease. Contrary to popular belief, most fish do not use their fins to swim. They force themselves forward by body movements and use the fins only as ailerons, rudders, and brakes. Swimming speeds vary greatly with different species. The swordfish, according to some observers, has been clocked at sixty miles per hour, while the sea horse needs a full five minutes to swim three feet.

All fish swim, but most people are surprised to learn that some species can also walk, climb, and fly. Some species walk about the ocean floor on sharp, downward-projecting fins that act as stilts. Another species, the mudskipper, is so active that, even if stranded on dry land, it can often outmaneuver a human pursuer. Still other types carry fins pronged in such a way that, searching for insects, they can climb trees at the water's edge. Ray-type fish have huge fanlike side extensions, which they flap in a flying motion as they swim beneath the surface of the water. Certain other fish have greatly enlarged pectoral, or "shoulder," fins, which they spread to soar through the air like a bird above the surface of the water.

Perhaps the most remarkable fish equipment is designed for depth control and equilibrium. With a small and fantastic gas factory, attached to a balloon-like organ called the swim, or air, bladder, the fish manufactures gas to inflate or deflate its balloon in order to increase or decrease its body weight. Without the ability to release the gas and thus increase its relative weight, a fish could not dive into deeper water where the pressure is greater. And, once down in the depths, it could not again rise to the surface without refilling its internal balloon to aid the process.

Never does a fish become confused, even in the darkest of waters, and swim upside down. It is provided with another small miracle, called the organ of equilibrium: small, loose, hard pebblelike objects, referred to as either otoliths, or statoliths, grow within a small hairlined chamber known as the statocyst. As the fish moves about, the downward pull of gravity controls the position of the otoliths. As the little pebbles press against certain of the hairs lining the cavity, their attached sensory nerves carry a message to the fish's brain, telling it whether it is upright or upside down.

Fish reproduce in fabulous numbers and, in the overall plan of nature, almost all their young are merely food for other water animals. An extremely small percentage ever grow to sufficient maturity to take advantage of the marvelous survival equipment with which the adult is endowed. The most skillful of artists could not duplicate the camouflage coloring with which many fish are protected. Many other fish species depend for survival upon extremely swift reflexes and speed. Still other fish have exceedingly sharp and vicious teeth. Some have sharp spines or poison glands as described more fully later in this chapter. A few species are provided with "batteries," and employ electric shock as a weapon of defense, while other, more passive, types, such as the pipefish, are masters at using their shapes in the art of deception and camouflage.

Now that we have briefly run through the marvelously intricate and subtle mechanisms of the fish, let us turn our attention to a few of the fascinating fish oddities that swim in the sea, discussing them in alphabetical order by common name.

Crucifix of the Sea

A cross suggesting the structure upon which Christ was crucified is a symbol of the Christian religion. It is not unusual to find crosses in the artistry of nature on mountains, trees, and so on, but nowhere has nature's handiwork been quite as amazing and complete as her crucifix in the skull of the common saltwater catfish found in the coastal waters of Florida (Plate 10d).

On the underside of the skull—actually the roof of the mouth (with all flesh cleaned away)—the bone is formed like the figure of Christ on the Cross. The halo and the crown are all in plain view. A gentle shake will rattle two pearly bones within the skull, said to be the devil's dice cast by the Roman soldiers for Christ's garments. Turn the skull over, and the figure of a priest, wearing a crown and robe hanging in long lacy folds, is revealed. His arms are outstretched in an attitude of prayer. The twelve ribs represent the twelve apostles. And the sharp-pointed notched bone in the dorsal fin is considered a replica of the Roman soldier's spear that pierced Christ in the side.

For centuries, the people of Latin America have looked upon this odd crucifix shell with reverence and awe, and have kept it in their homes as protection against illness and injury.

The two pearllike bones in the catfish's head (the devil's dice) are actually the otoliths, which are used as a balancing mechanism to aid in locomotion. At one time these curious pearllike formations played an important role in the lives of superstitious people. Many people carried an otolith at all times to ward off the evil eye and as a cure for colic.

This special fish is also noted for the oddity of its reproduction. The female lays the largest eggs in the fish world, while the male scoops them into his mouth to carry them until they hatch. He eats nothing at all during the nine-week incubation period and his body grows very thin. Then, during the remainder of the year, he eats constantly, growing fat and vigorous.

Eel

The long slender eel, which is actually a fish, is one of the many dangers at sea that a diver must face as he collects coral and seashells. Several different species of this odd animal exist——some of which, though not all, are quite vicious and dangerous—and they thrive in abundance on the coral reefs of the world as well as along our Atlantic coast and the European coast. The eel is one of the earth's most fascinating and mysterious creatures. So unusual is its life cycle that much still remains to be learned after many years of observation and research.

In 1856 Dr. J. J. Kaup, of the Paris Museum of Natural History, found a small three-inch fish, which he identified as *Leptocephalus brevirostris* (or leaf fish) because it resembled a veined willow leaf. Forty years later, in 1896, two Italian biologists were eyewitnesses to a bodily transformation of the leaf fish, and science then knew the leaf fish was but the larval form of the eel.

Finally, in 1922, after an eighteen-year search, Johannes Schmidt, a noted Danish marine biologist, discovered the birthplace of the eel. He traced the leaf fish thousands of miles out into the ocean, finding smaller and smaller leaf fish until he finally determined their source to be the Sargasso Sea about midway between Bermuda and Puerto Rico in the North Atlantic Ocean. Although he was successful in finding the young larvae, he failed in his attempt to locate the mature female with fully developed eggs.

A 1966 Danish expedition to the Sargasso Sea, led by Dr. E. Bertelsen, also trawled up eel larvae in an early stage of development, but it, too, was unable to find a breeding mother. Though many questions still remain unanswered, marine biologists have by now made most interesting progress in unraveling the story of the eel. Though part of its life cycle takes place in the deep sea where it cannot be observed, the eel is believed to go through more body transformations than any other vertebrate animal.

The Sargasso Sea may, it is thought, be the spawning ground for both the American and European eel. They are believed to lay their eggs from January to March beneath heavy seaweed growth at a depth of from twelve hundred to fifteen hundred feet.

Tiny, transparent, ribbonlike larvae hatch from pinhead-size eggs. Perhaps as many as ten million are produced by one female. Billions of these living ribbons are carried northward by the currents of the Gulf Stream. With no self-propelling force, their course is entirely at the mercy of the elements. Great quantities are preyed upon by fish and jellyfish as they continue their journey—and their growth. A little later they gain the ability to steer their own course and their clear bodies become cloudy. By some miraculous instinct, the American and European species then sort themselves out according to ancestry. With a God-given ability reaching far beyond the understanding of man, they then use a remarkable homing instinct and swim back to the home shores of their parents. The eel is known to have a series of extraordinary taste buds in a band along each side of its body, and it is believed these extremely sensitive organs of taste may play some role in its ability to unerringly find its way across those thousands of miles of intervening water. The young of our American species require a year to work their way back to our shores. The distance being greater, it takes the European species three years to complete their amazing journey.

By the time they reach their respective coasts, the tiny, formerly transparent ribbons have grown to a length of 2½ to 3 inches, and have assumed a cylindrical, opaque form like the adult. While growing very slowly, they then, for a few months, haunt bottoms of pools and channels along the coast, eating anything they can find—animal or vegetable, dead or alive. At this stage they are the immature yellow eel, active mainly in summer, so often seen and caught in shallow coastal waters.

The females of the small yellow eel then find their way into freshwater creeks, rivers, lakes, ponds, and ditches. They spend the next five to twenty years swimming upstream, going farther and farther inland as they eat and grow to a possible forty inches in length. Rivers especially noted for this female migration are the Potomac and Mississippi in the United States, the Waal in the Netherlands, and the Severn in England.

During this migration, the males remain in the ocean where they grow only to an approximate length of twenty-four inches.

Complete mystery shrouds the motivational force that brings about the return of the female migration to the sea. For apparently no reason at all, the females will choose a dark autumn night, and just suddenly turn and start swimming downstream to rejoin the males waiting in the ocean. As they swim downstream, still another physical change takes place. Their bodies become a glowing silver color, and the blunt rounded nose becomes pointed.

Reunited with the males, they go swiftly and unerringly together (and it is believed without further food) through the thousands of miles of deep ocean to the Sargasso Sea. There, in the tropical waters of their birth, they spawn and die, their life cycle completed.

The eel is now playing an important part in marine medical research. The large electricity-producing organ of the electric eel, which can give the unwary swimmer a severe electric shock, contains an enzyme from which an effective pesticide antidote has been made. It is now believed this substance may prove very helpful in treating many nerve-related human illnesses such as various kinds of emotional disturbance, muscular dystrophy, and epilepsy.

The eel also serves as an excellent food fish, and great quantities are taken annually for the markets of the world, but the dangerous species, such as the moray, are a mixed blessing. The moray's living and feeding habits make it a menace to swimmers and to all ocean life. It chooses crevices in and around rocks and coral for its home, backing into them with only its head and teeth exposed. Then, when any animal, including man, swims by, it quickly attacks and greedily tears away bites of flesh to eat. Conversely, however, there are other eels that operate out of a principle of self-protection rather than attack. One curious species, the snake eel, has the interesting faculty of being able to reverse direction and swim backward. It can also retreat into the sand, tail first, by twisting its body down into the sand with the sharp point of its tail, just as if it were a corkscrew, auger, or brace drill.

Globelike Fish

The globefish, burrfish, porcupine fish, and puffball fish are saltwater fish of the Diodontidae family. Each has an abdominal wall capable of great expansion, and some species can fill their bodies with air to the size of a basketball, making it very difficult for an enemy to swallow them. When extended, they look like large spine-covered globes or balls with a beak and tail attached (Figure 5).

I have been told by some observers that if a large enemy appears on the

Figure 5. Puffball fish in a state of expansion. *J. W. LaTourrette, courtesy Wometco Miami Seaquarium.*

scene, for still greater protection, some species assemble in large clusters with their spines intermeshing. Thus, they present a united front, and even small fish in the school are safe from the largest of the enemies. I do not state this as a fact for I know these observations and conclusions are not as yet accepted by all. To me, it would seem a most logical thing for this type of fish to do.

Fish of the puffer (blowfish) family are often highly poisonous, depending upon the species, locality, season, and the skill with which they are dressed for eating. Indeed, in Japan, one common method of suicide is to eat a certain Pacific species, for which no antidote is known, that is so highly toxic as to cause death in a matter of minutes. Properly chosen and skillfully cleaned, however, puffers can be eaten safely. But the incidence of death from contaminated flesh is so high in Japan that the government provides special puffer "cleaners."

Pipe, or Pencil, Fish

The long slender pipefish (*Syngnathus griseo* Linnaeus) belongs to the same family as the sea horse. It lives along the shore, and swims among eelgrass in search of tiny crabs and miniature sea animals. When frightened by an enemy, the pipefish hastens to a bed of eelgrass, and stands on end at the exact angle the grass is bending in the moving water. So remarkable is its resemblance to the grass, that it seems to simply vanish from the sight of its enemy.

In the pipefish nature perfected the first zipper—many centuries before man awakened to its use. The male pipefish has a long slender groove almost the full length of his abdomen. Into this groove the female places her fertilized eggs. A thin membrane then zips over the top, holding them in place through the sixteen-day incubation period. Then nature unzips the pouch to release the one-half-inch-long infants.

Rays

Rays, or Batoidea, are so far from the fish norm that it is hard to even think of them as fish. Shaped like flat pancakes, they move about on the ocean bottom in search of crabs, oysters, and other animals for food. Their thin bodies taper to almost nothing at the edge, and they seem to be flying as they flap their big fins through the water.

The largest and most powerful of the rays is the sea devil, or *Manta*. It reaches a weight of twelve hundred pounds and a width of twenty feet and, when harpooned, will drag a boat with great swiftness. It feeds on plankton and schools of shrimp and small fish. Because of its terrifying appearance, it is greatly feared by pearl fishermen and others who work beneath the sea. However, it is not actually as dangerous as it looks for its teeth are small and its tail is spineless.

The real sea devil is the twelve-foot-long stingray (Dasyatidae), common along our Atlantic coast, which is one of the most dangerous and most feared of the rays (Plate 11a). Its slender whiplike tail is armed with a long poisoned spine, set with backward-projecting teeth. Used as a weapon, these teeth make deep lacerating wounds. In man, such a wound is very liable to result in blood poisoning.

The stingray inhabits nearly all seas, lying flat on the sandy bottom, eating crabs, oysters, and other mollusks. Its eggs are hatched within the body.

The skate (*Raja*), which is extremely flat and wider than it is long, grows to an average size of about eighteen inches. There are approximately forty known species of skate, of varying shapes and sizes.

The skate's method of reproduction is of special interest. After one mating, the female lays two fertile eggs about every fourth day during a nine-month season. By instinct, it swims to a shallow warm area with a bottom weed covering and a food supply just right for infant skates. There it lays a rectangular, black, leathery egg case, enchantingly known as the mermaid's purse. A tiny baby is growing inside the case, and projections at each of the four corners firmly attach it to the weeds. Thus, when the baby hatches, with its wings folded like a pancake, it is sure to find itself in the proper environment for survival.

The mother returns to sea, leaving the case unattended. It remains attached to the weeds for a full year before hatching. After it gains sufficient size and strength to do so, the baby leaves the sheltered area of its birth and swims out to sea to continue the life cycle.

The infant skate is truly a charming little creature. I have two in my collection smaller than the palm of my hand. Viewed from the underside, they look for all the world like small human faces with mouths spread in broad grins.

The torpedo (*Narcobatidae*) is a remarkable family of rays. Its special distinction is an electric organ at the back of its head, which can benumb an enemy with a severe electric shock. This ability has also given it the names crampfish and numbfish. It is found in all warm seas, though not in abundance. One of the largest species known, the *Torpedo nobiliana,* is found along our Atlantic coast. It grows to a five-foot length and sometimes weighs two hundred pounds.

At one time the shock from an electric ray was believed to be a cure for gout. It is no longer used because live charged rays were hard to obtain, and also, partly because most patients preferred the gout.

Reef Fish

One of the greatest charms of the Great Barrier Reef of Australia is its multiple array of brightly colored small fish. I found them most fascinating to watch.

By day, the beautiful grunters and snappers swim in protective schools and contribute much to the total coloration of the reef. At night they scatter over a wide area to feed. As they leave the protective school, the bright colors fade and they assume a camouflage coloration pattern to match the various feeding areas. The bizarre, highly colored grunt fish is considered the comedian of the fish world. It crawls over the rocks on its long fins using them as fingers, and turns its head slowly from side to side. Then it rolls its eyes in such a way as to leave a person feeling he is being winked at. When removed from the water, it makes a grunting sound from which it gets its name.

A few other fantastically colored reef fish are the parrot, angel, surgeon, and damsel—each with its own special pattern of behavior.

Sawfish

Another of nature's oddities is the peculiar sawfish with its long, flat, tooth-edged bill, or saw, which extends one-third the length of his fifteen-foot body. Florida's large sawfish has been given the name *Pristis pectinatus*.

Sawfish inhabit the mouths of rivers where they wreak havoc among schools of mullet—their favorite food. Their slot-like mouths, which are on the underneath side, are incapable of catching moving fish. So they lie on the bottom, with their mouths flat on the ground, breathing through holes in the tops of their heads. When a school of fish appears, the breathing holes in the sawfish's head close, and it breathes through its mouth as it rises to unmercifully slash the school of fish with its sharp-edged bill. Later, at its leisure, it eats the mangled fish from the ocean floor where they settle.

The young are born alive with a velvet-like glove fitted over the saw to protect the mother. As many as twenty-five may be born at one time.

The sawfish does not saw through a boat as is sometimes believed, but a large one can slash a man in half. If one is hooked from a boat, its strength is sufficient to tow the boat for many miles.

Sawfish skin can be made into tough leather. Its liver yields valuable oil, and in China its fins are made into soup.

Sea Horse

The sea horse (*Hippocampus hudsonius*) is an odd little creature with a colt's head, an insect shell-like body, and a kangaroo pouch. It is actually a form of pipefish only five to eight inches long, one of fifty known species of that type of long slender fish. Its little head and arched neck look for all the world like a prancing pony.

Almost impossible to distinguish, it lives upright in shallow water with its tail wound about seaweed stems (Figure 6). Thousands live in the shallows near Key Largo, Florida, but quickly disappear if the water becomes either too salty or too fresh. One little fan-shaped fin on its back is its only means of locomotion. And the little eyes swivel independently, looking both forward and to the rear. Often during storms the seaweeds break off and carry great numbers of sea horses out to sea.

Caught up on the Gulf Stream, it is not unusual for little Florida sea-horse voyagers to ride eastward to Europe or into the Sargasso Sea. Throughout their journey, bountiful nature provides them with a feast of other small creatures, eggs, embryos, and plankton, that are also riding the Gulf Stream. When they eat, a trap door at the end of the snout opens, and the passing food is sucked in with a quick intake of water.

The female deposits her eggs in an abdominal pouch of the male. He fertilizes them as they are being deposited, and carries them throughout a sixty-day incubation period. He then gives birth to the baby horses—a few at a time. According to some authorities, one birth may produce as many as four hundred babies.

Closely related to the sea horse is the tiny pigmy horse, *Hippocampus zosterae* (meaning horse-caterpillar). This lovely tiny creature, no longer than a paper clip,

Figure 6. Sea horses among natural grasses. *J. W. LaTourrette, courtesy* Wometco *Miami Seaquarium*

takes on the coloration of its environment and becomes beautiful shades of brown, green, yellow, and so on, which, of course, makes it very nearly impossible to see as it clings to its weed or sea-grass blade. This camouflage is its main defense, since they are extremely slow swimmers.

Museums and home aquariums create a huge demand for these little "horses," which are fished for by trawlers that can often capture two thousand per day in nylon nets.

Stonefish

Nature often goes to extremes in creating ugliness as well as in creating beauty. Certainly the stonefish (*Synanceja trachynis*) is among the very ugliest, as well as the most dangerous, of her creations.

Among the many perils of the Great Barrier Reef, the stonefish bears the distinction of being the most dangerous; not that its venom is more deadly than other poison life on the reef, but because its camouflage is so fantastically perfect. With learned care, the other dangers can be avoided, while the stonefish cannot be avoided because it cannot be seen.

It grows to a length of one foot with its scaleless body covered by an ugly warty skin. In both coloring and form it is so exactly like the eroded limestone among which it lies that no one can see it. It is even coated with the same slime and seaweed fragments found on the stones. Its mouth opens upward from the top of its hard bony head. And the eyes are placed in the bottom of two deep depressions. Along its back is a row of thirteen ivory-tipped spines that barely show above the puckered skin of its back fin. Each of the spines is provided with a sac of poison venom.

The stonefish is a very poor swimmer. So, as danger approaches, by instinct it freezes into total immobility and waits to be stepped upon. With the least pressure on its back, the sharp spines dart up with enough force to pierce a shoe sole and eject their poison. Even the slightest puncture, or merely a skin wound, is sufficient to cause intense agony that can last for weeks.

The Australian aborigines perform a strange dance that has been practiced down through the centuries. Their forefathers, knowing the great danger of the stonefish, used the dance to impress upon the children the great and terrible agony that would befall them if they stepped upon a stonefish. In performing the dance, the dancer impersonates a beach walker, carefully watching his footing as he searches through the tide pools. Then, suddenly he screams as he limps to shore holding a clay model of the stonefish in his hand. He then throws himself on the ground as he writhes and screams in agony, while the children are shown the thirteen spines that caused his distress. The witch doctor is then called in to dance about the victim and perform his magic. Failing to bring relief to the patient, he throws his hands up in despair. The tribesmen then wail the death chant, as the children watch the dancer "die" of his terrible wounds.

7. Giants of the Sea

Our sea story would hardly be complete if we limited our discussion to the smaller animals of the sea; so let us now round out the picture with three of the ocean's huge giants: the whale, the whale shark, and the walrus—enormous creatures that have intrigued and served mankind down through the ages.

Whale

The mighty whale has been a figure of fascination for centuries. Herman Melville's masterpiece *Moby Dick* is probably the best-known work to deal with the whale as an animal of mystery and symbolic meaning, but it is certainly not the only one. The whale has figured in myth and legend from earliest times, probably partly because of its great elusiveness. Though one may see an occasional whale washed ashore, usually dead, or else reduced to helplessness out of its watery element, the casual whale-watcher must ordinarily be content with a brief partial glimpse of the animal as it swims or comes up for air. The whale, which belongs to the order Cetacea, along with the dolphin and porpoise, is one of the largest known living animals.

Although it lives in the sea, the whale is a warm-blooded, air-breathing, true mammal. They bring forth their young alive, and suckle them in the same manner as land mammals. To keep them warm, the all-wise Creator provided adult whales with a two-foot-thick, thirty-ton blanket of blubber—the perfect insulation material. Not only does it preserve warmth, it reduces the specific gravity of their mighty bulk to make swimming possible. And, to enable them to breathe in their watery home, their grotesque nostrils were created to turn upward with the openings, or blowholes, in the very top of their huge heads. For each breath of air the whale must open the blowholes as it raises the top of its head above the surface of the water. Then it quickly activates its watertight seal so it won't drown as it again proceeds with its underwater life. To give it the immense driving force needed for swimming and fighting, its tail was created to move horizontally instead of vertically as in true fish.

The food of many whales consists chiefly of the small mollusks and crustaceans that are so abundant in their immense sea pastures. The toothless baleen whale eats nothing but plankton. It is equipped with a huge straining apparatus, called baleen, or whalebone, through which it strains enormous quantities of minute planktonic plants and animals from the water. Some species grow to a length of 95 feet and weigh 150 tons. A single specimen has been known to yield 3,500 pounds of whalebone, and it is of great commercial importance. Whales in the sperm group have teeth in the lower jaw and live almost entirely upon squid. They usually feed on small squid, but they are known to lock in battle with the giant species of the North Atlantic—these giant squid (*Architeuthis*) reach a fifty-five-foot length and have thirty-five-foot arms with huge suction disks that sometimes leave large circular scars on the attacking whale.

The whales of the orca group are the only ones that attack and devour warm-blooded mammals. Known as killer whales because they are one of the few animals to kill more than they can eat, their many teeth are most impressive and of enormous size. They can dive to great depths and remain under water for extended periods. Moreover, with their extra-large eyes and keen vision, they can also see well in the air above the water surface. These twenty-foot killers fear nothing in the sea and even attack and eat the large, and very vicious, California gray whale. Because of their speed, strength, and extreme ferocity, they are called the tigers of the sea. Strangely, in spite of their ferocious nature, in captivity they are marvelously receptive to training, and develop a close warm friendship with their trainer. Their intelligence is greater than that of the dolphin, and it is hoped the killer whale may someday provide a means of communication between animal and man.

The fifteen-foot beluga whale of the far north has such a huge appetite for salmon that it poses a threat to the salmon industry. Biologists from the Alaska Department of Fish and Game are now making some headway in solving the problem. Taking advantage of the beluga's intense fear of its mortal enemy, the killer whale, scientists are now experimenting with tape-recorded sounds of the killer whale with the hope of learning how to frighten the beluga from the areas of greatest salmon concentration. The clever biologists broadcast their recorded sounds under water from the cannery dock two miles upstream from the mouth of the Naknek River. During a three-day test period, they saw as many as twenty-five belugas swimming upstream for their salmon feast. And, when they turned on the killer whale sound waves, the belugas quickly dove from sight and were next seen swimming back out to sea.

Of all whales, perhaps the sperm whale (*Physeter*) has the greatest commercial value because of the superior quality of its oil. It is an enormous animal, found in all warm seas, with some species reaching sixty feet in length. Before the introduction of petroleum oil and electric lights, large fleets of whaling vessels sailed to all parts of the world to supply the oil needed for candlemaking, soapmaking, and so on. Then, for a time, demand for the sperm whale lessened, until high-speed motors were invented. It was learned that, for these motors that operate machinery running at high speeds and temperatures, sperm oil surpassed all others, and was considered of critical importance. It became so vital as a lubricant in transmissions, machine tooling, and rifling guns that the United States Office of Emergency Preparedness began stockpiling it. So, once again, whaling vessels scattered over the globe to satisfy an even greater demand. So many sperm whales were captured for their oil that in 1970 the sperm whale was placed on the list of endangered species.

To save the species, an urgent but unsuccessful search was made for another comparable natural oil. There were also many attempts to artificially produce a synthetic oil of like properties, but all experiments failed. Then at long last it was discovered that the peanut-sized seed of a small desert shrub, the jojoba, contained a liquid wax remarkably similar to sperm oil. The bush is native to a 100,000-square-mile area of the Sonoran Desert of Arizona, California, and Mexico. The Indian Division of the Office of Economic Opportunity made a grant of $100,000 and, using Indian labor, it is planned to harvest a tank carload of the jojoba oil for industrial experiment.

Though a successful substitute oil would lessen the demand for the sperm whale on this front, the animal would still be hunted and in danger of extinction because of its spermaceti and ambergris, for which there are no substitutes. Spermaceti, a waxy substance prepared from a special oil found in the head of the sperm whale, is very valuable for candlemaking. Ambergris is another waxy substance secreted in the intestines of the sperm whale, which is used in the preparation of costly perfumes.

The great sperm whale has also enriched the world of both legend and art. It is the great sea beast of *Moby Dick,* and it is the ivorylike teeth of this huge beast that inspired the ancient art of scrimshaw.

Scrimshaw came into being during what historians call the great age of whaling and sailing, from 1770 through 1900. The term *scrimshaw* refers exclusively to carving, or etching, executed upon whale teeth, whalebone, and walrus ivory by New Bedford whaling men.

Those long whaling voyages, contrary to what one might think, were not all excitement. Often, after a catch, there was a dull deadly monotony of several weeks duration before the next sighting. To while away the long boring days, the sailors began scratching designs on whale teeth. To emphasize the design, they would rub some dark substance into the etching—just anything handy: tar, lampblack, soot mixed with oil, or even tobacco juice. No one knows who actually started it, but, as the years passed, both skill and tools improved until the time-killing pastime grew into real art. At the peak of the whaling era, it was estimated that 200,000 seamen were etching, whittling, and carving to while away the endless hours between catches.

Like so many folk arts, scrimshaw over the past sixty years has almost disappeared. With the departure of sailing ships, and the advent of modern technology, the whaling man no longer finds himself with either the spare time or the inclination to indulge in the painstaking pastime of scrimshaw.

This now almost forgotten art is still being kept alive by a few tireless artists, and it is accepted as one of the most important true American folk arts—although most of the work was executed while the sailors roamed far-off seas. The carvings become increasingly hard to find and now often sell for amazingly high prices for a single tooth. I feel quite fortunate to have found a lovely one to help enhance my collection.

Whale Shark

Our second giant of the sea is the whale shark (*Rhincodon typus*). The whale shark is a shark—a true fish—not a mammal as is often supposed because of its resemblance to the whale. The largest of all fish, it grows to a forty-five-foot length in the warm waters of tropical seas. It is equipped with gill rakers, which it skillfully employs to collect plankton food. Only occasionally does one venture into shallow waters, where it is then usually either stranded or harpooned.

Fortunate indeed is the nature lover who is privileged to see one of these monsters. Its huge gaping mouth, frightening to swimmers who encounter one, has won for it the name monster shark. Actually, however, it is slow-swimming and toothless, with a gentle, fearless, docile nature. This mammoth frightening creature is innocent of harmful intent as it swims about through the tropical currents of the oceans in search of its plankton food. However, it does constitute a hazard to small boats. Since it has no fear of a boat or motor noise, small-boat occupants are in danger of being capsized as one of the huge forty- or fifty-foot-long monsters rises to the surface beneath them.

Each time a whale shark is sighted, so much excitement follows that the story grows and grows, until one is never certain where fact ends and myth begins. However, the year 1912 brought proof to the doubters, when a thirty-foot-long whale shark with a girth of eighteen feet was put on exhibition in Miami. Captain Charles Thompson, after an eight-hour struggle, had been able to harpoon and strand it. A number of these whale sharks have been sighted in the warm currents along the eastern shore of Australia where the water is especially rich in plankton during certain seasons of the year. To make it authentic, one has now been photographed there.

The whale shark's hard solid body is covered with rough coarse skin of a sandpaper texture. It is known to make a constant croaking sound that is believed to be a kind of sonar device used in navigation. I have in my collection a huge mermaid's purse on the order of the skate egg case, but many times larger. It had washed ashore on the southern tip of

Africa. I have been told it is the egg case of the whale shark, although I am waiting for more authentic confirmation.

Walrus

To most people the walrus, with its long moustaches, is a figure of fun, but this animal is in fact one of the most interesting and valuable sea creatures. The huge walrus is indeed a glorious spectacle as he lies on the ice floes and islands of the Northern Pacific, off Alaska (Figure 7). The walrus (*Odobenus rosmarus divergens*) grows to a length of twelve feet, attaining a weight of three thousand pounds. It supplied all needs for the early Eskimo, just as the American buffalo did for the early plains Indian: its flesh and its stomach contents provided the Eskimos with food; their summer homes and boats were covered with its hides, which also supplied leather for the dog harnesses and fishing lines; cash was acquired by the sale of the ivory teeth and tusks; the tusks were shaped into the vital spears and harpoon points; and the intestines were slit and made into rain parkas.

In the early days the walrus supply kept pace with the Eskimos' demands. Then, in the mid-nineteenth century New England whalers began, when whaling became poor, to harpoon the walrus for its oil and ivory. Next, both the whalers and the Eskimos substituted rifles for their harpoons and real trouble began for the walrus. A recent United States government report states that in one year the Eskimos killed over three thousand walrus with their rifles and let more than half of these sink to the bottom or swim away to die, unable, with the rifle method of killing, to salvage them for use. The ivory-hunting white men, who killed merely for the walrus tusks and discarded the rest of the animal, perpetrated even greater ravages.

At the time the rifle was substituted for the harpoon, it was estimated that the Pacific walrus numbered two hundred thousand. Now there are reported to be only about forty-five thousand left. The walrus is a slow breeder, and estimates say that, for the birth of every calf, two adults are being killed. Extinction is now a certainty unless the government interferes.

The state of Alaska has now passed laws hoping to regulate the kill limit and the ivory traffic, but their knowledge of the life habits of the species is so limited they are fearful of the final outcome. Research projects are now in progress, but gathering the facts on the walrus is proving to be a quite awesome challenge for scientists. They need to know more about its reproduction, migration, food needs, natural enemies, and so on before they can hope to lend a hand in restoring it to its original numbers.

Researchers now believe the walrus's major food is shellfish. At a depth of 180 to 300 feet, the walrus digs the shellfish from the ocean floor with its tusks. And scientists believe a large walrus must have the equivalent of one hundred pounds of food daily to survive. There is evidence, however, that a walrus may fast for a considerable time and then consume tremendous quantities of food on the days it does eat. It is believed the walrus also

Figure 7. A bull walrus, photographed off Togiak Bay, Alaska. *William Sholes, courtesy U.S. Fish and Wildlife Service*

prey occasionally upon seals. The position of the tusks makes it impossible for the walrus to tear off chunks of solid food. But it is equipped with odd protruding pads on each side of the muzzle with which to force food into its mouth. The pads serve as both lips and hands and are the total tools for food intake.

Newborn walrus calves are usually about four feet long and weigh from 100 to 150 pounds. They are born from April through June, on the ice floes, after a thirteen-month gestation period. They are immediately able to slip into the water and swim away with the mother. She nurses the baby from twelve to eighteen months, but continues traveling by its side several months after weaning. The calves grow very rapidly and reach maturity in from four to six years. The female matures somewhat earlier than the male, and reaches about two-thirds its size.

It is believed the walrus has few natural enemies other than man and the killer whales. I did not personally see it happen, but when I was in the Bering Sea area I was told of Eskimos who saw killer whales throw a large two-thousand-pound walrus completely out of the water. The walrus is deathly afraid of the killer, and a whole herd will panic when one appears in their midst. In 1936 a famous panic occurred on Saint Lawrence Island some distance off the West Alaska coast in the Bering Sea. When attacked by a killer, a huge herd, frantic with fear, plunged from the water onto the beach in one huge pile, crushing and smothering those on the bottom. Later, two hundred head were counted, dead, on the beach.

8. On the Beach

There are many things to enjoy in a trip to the beach: sunning and swimming, of course, collecting seashells, and so on, but there are also many other things of interest to the nature lover. Now we will talk about just three of the attractions—turtles, beach birds, and good-luck-charm beach treasures—that make beach going such a fascinating experience.

Turtle

There are some three hundred known species of the ungainly turtle, a reptile (along with snakes and lizards) belonging to the order Chelonia. Properly speaking, the term *turtle* should only be used for the marine, or ocean, species. The land-based species are correctly referred to as tortoise, and the term *terrapin* is reserved for those found in freshwater. However, since the word *turtle* is popularly used for all three branches, we shall do the same. In any case, this interesting animal cannot be understood without considering all three forms, so, although this nature story is fundamentally of the sea, it will be essential in this case to focus inland for a time to make our story complete.

All three forms of the turtle are encased in a protective shell; a top covering is called the carapace, and the lower plate, the plastron. The land tortoise can retreat in its entirety between its shell plates for protection, but the sea animal is unable to withdraw its head into its shell. Its feet also differ, as they are equipped with seal-like flippers for swimming. Turtles do not have teeth, but their jaws are lined with heavy, chisel-like biting edges. Most species are carnivorous, although some eat vegetation. They all lay eggs covered with a leathery shell, which most species bury in the sand to be hatched by heat from the sun.

Marine Turtle

The sea turtle is a most intriguing animal. It is truly enthralling to watch the big females construct their nests. A massive turtle form emerges from the water and goes lumbering awkwardly over the sloping sands of the beach. Strong sweeps of the large flipperlike feet propel the big body forward, leaving broad tracks in the sand. At frequent intervals the poor thing settles down with an audible sigh to rest for a while. This sighing, and the scraping of the heavy body over the sand, can be heard at a distance of several yards. Her awkward struggle continues until she reaches the very top edge of the beach where vegetation begins.

Reaching her destination, which by some strange instinct she recognizes as exactly the right spot, she begins the construction of her nest. With backward sweeps of the strong front flippers and a rotation of her body, a hole soon begins to emerge. The strenuous effort, alternated with more sighing and resting, continues until the excavation is as deep as her body. Then, with an uncanny ability and precision, she digs a shaft down into the moist sand beneath the center of her excavation. To make this vertical shaft, into which she later drops her eggs, she uses her stubby rear flippers with the skill of human hands. With the cupped end of a flipper, she scoops a handful at a time of the damp sand and flips it aside, all the while using the broad side of a flipper to firm the sidewalls and to prevent their collapse. With the shaft excavated as deep as she can reach with the flippers, she then drops up to two hundred eggs down into it.

Turtles are known to leave the sea as many as seven times in one egg-laying season, but they usually lay fewer eggs their first trips out—more nearly fifty than two hundred. After the eggs are dropped in the shaft, she very carefully covers them with sand and pats it down firmly on top. She then rakes in dry sand from the surrounding area to fill, and conceal, her large excavation. The task completed, she wearily lumbers her way back to the sea with still more sighing. Going to and from the nest she is apprehensive and fearful of human presence, but during the time she is busily occupied with the actual building of the nest, she does not seem to mind a close human observer.

After nine to ten weeks of warmth from the sun, that shaft full of eggs miraculously becomes a teeming center of life. Up to two hundred infant turtles awaken to find themselves buried beneath a foot or more of sand. Again, by a provision of the great Creator too intricate for human understanding, these tiny two- to three-inch babies survive with only instinct to guide them. They struggle upward through that thick heavy layer of sand and unerringly head for the sea—never toward land. They scuttle across that intervening sandy beach with a speed and agility more usually associated with the hare than with the tortoise. But, even so, that short journey to the sea is fatal to many of the little animals, who are captured by beach predators, crabs, and birds. Once in the sea, they seek nooks of comparative safety but, even there, so many are eaten by enemies that but a small percentage survive to reach adulthood.

There are but five known species of sea turtles, and they are widely distributed in all the warm seas of the world. Four of the species are quite common:

The leatherback, or trunk, turtle (*Dermochelys coriacea*) is the largest known turtle, reaching eight feet in length and weighing eight hundred to twelve hundred pounds. Its flesh has a very unpleasant taste, and I am told it can cause sickness if eaten.

The loggerhead (*Caretta caretta*) sometimes weighs up to 450 pounds. It is captured chiefly for its oil, although its flesh, also, is often sold in the markets. The loggerhead female lays a thousand eggs at one time.

The hawksbill (*Eretmochelys imbricata*) is the smallest of the sea turtles, rarely exceeding two and one-half feet. It is distinguished by its sharply curved beak and extra-beautiful shell. This shell, the popular tortoiseshell, takes a beautiful polish and is of high commercial value. The hawkbill is found in Atlantic waters from North Carolina to Brazil.

The green turtle (*Chelonia mydas*), which grows to a length of four feet or more, is probably the most valuable reptile in the world (Plate 11c). Adults usually range between two hundred and three hundred pounds, though often they grow much larger. The largest big green ever recorded weighed 850 pounds.

This large highly edible turtle was an important factor in the early colonization of the American tropics. Its great abundance and the fact that it could be kept alive for months with a minimum of care made it an ideal food solution for the early British and Spanish explorers of the New World.

Uses for the green turtle have multiplied and demands increased down through the years until it has been greatly feared the species would be hunted to extinction. There is now an ever-increasing demand for frozen turtle meat, and the popularity of the clear green turtle soup takes a heavy toll of the population. The delicious soup is made from a three- to four-pound section of cartilaginous material called calipee, which is found beneath the shell of the green turtle. Its ever-increasing demand has excited opportunists to a new frenzy of calipee hunting. Eager for the high price they can get for the little strip of calipee, they leave many huge turtles to die after removing this bit of treasure from beneath its shell.

An increasing trade in turtle hides for leather goods is also adding to the threat of extinction. In fashionable New York stores catering to the wealthy, they sell turtle-skin objects as status symbols. Only a small portion of the turtle's skin will do, and it brings such high prices that many turtles are now losing their lives to skin hunters. Taxidermy also takes a heavy toll for there is a great demand for the stuffed animal in museums and exhibits.

It is believed there were many green turtle rookeries in the Western Caribbean during early colonial days. But now they are mostly hatched on a single twenty-two-mile beach strip along the eastern coast of Costa Rica in Central America. Since 1954 this short beach strip has been the center of operations for scientific research on the problems involved in restoring the green turtle to its former abundance. The actors in the drama are the Costa Rican government, the National Science Foundation, the Caribbean Conservation Corporation, the University of Florida, and even the United States navy gets into the act as it ferries infant turtles hither and yon throughout the Caribbean area for experimental purposes.

So far, the results of these experiments are unknown, and no marked success in increasing the green turtle tribe has been observed yet. Even their basic life cycle is still shadowy. It is known that every two or three years the female green turtle lumbers up on a sandy beach in clear view to bury a hundred or so golf-ball-sized eggs in the warm sand. (The male never comes ashore to share the many beach dangers.) She then disappears into the ocean and all efforts to solve exactly how or where she migrates end mostly in frustration. By tag retrieving, it has been learned that she makes journeys of at least fifteen hundred miles and that she returns to home base when egg-laying time comes again.

We still do not know where or how the babies spend the first year after they crawl into the water from the sandy-beach nest. Their food during that stage also remains a mystery. Later, when they are four or five pounds in size, and possibly a year or more old, their jaws can cut turtle grass. They then go in schools over the subterranean pasture fields foraging in the manner of a herd of cows.

Since 1958 a green turtle hatchery has been operated on a section of the Costa Rican rookery. Nests are circled with fine wire net to protect them from predators and to capture the newly hatched babies. Many thousands of the babies have now been moved to many different beach areas where it is believed ancient rookeries existed. These infants are being "planted" in Central and South America, Mexico, Texas, Florida, the West Indies, and the Bahamas. Though no one knows yet the number of years needed for maturing, it is devoutly hoped that, when they do mature, they will return to their "foster" infant home to establish new rookeries, and thus ensure that our rapidly growing turtle needs will be supplied on into the future.

Tortoise

The problems of the tortoise, or land turtle, are also many and varied. We will briefly discuss here two species of the land form to more properly portray the full range of our turtle, which is certainly one of the most bizarre creatures in all of nature. Unlike the sea turtle with its flipperlike swimming feet, the land form has legs for walking. And, for greater protection against predatory animals, most land species have a higher domelike top shell.

In Africa there is one unusual species with another means of protection. It has a very thin flat shell and lives in a rocky area. When danger approaches, it crawls into a small crevice between rocks. It then tightly inflates its lungs to increase its size, until it becomes so tightly wedged in place that no animal can dislodge it.

The largest known land tortoise (*Geochelone elephantopus*, formerly *Testudo vicina*) reaches a giant size of several hundred pounds and lives over a century.

It is found on the small volcanic Galápagos Islands off the northwest coast of South America. Like many other unique animals, the huge Galápagos tortoise is fighting a losing battle for survival. For four hundred years, pirate, whaler, sealer, and merchant fleets have been stopping at the Galápagos to provision their ships with its live meat. The large tortoise can live in the hold of a ship for months with no food or water, and often two or three hundred are stowed away on a single ship at one time.

Also, their fat produces a valuable oil, for which huge numbers have been slaughtered. Moreover, they must compete for the limited food supply with the overpopulation of wild goats and donkeys now roaming over the dry islands; and rats and wild pigs take a heavy toll of the eggs and newly hatched young; and many times, for want of rain, the earth becomes so hardened above the eggs buried by the mother tortoise, that the newly hatched babies cannot escape and must perish in the nest. However, conservation authorities have now stepped in, with the hope of preventing extinction.

Terrapin

The terrapin, or freshwater, turtle is referred to by a variety of different names, such as diamondback, mud, painted, or slider turtle, depending on the species. Many of the species have beautifully colored markings, and many are highly edible. They constitute an important source of food, especially in the low areas along our eastern coast.

Beach Birds

Because of their special diet and habits beach birds play an important role in the overall ecology of the seashore. They are so plentiful and universal as to be near constant companions to any beach explorer (Plate 12a). The two species discussed below I personally found so especially delightful that it would not be possible for me to present them in any other way than to show them to you through the eyes of a shell collector.

The lovely little sandpipers are such a joy to watch and listen to. They travel in droves and always seem to be running at a merry pace just out of your reach as you walk down the beach; they are often so very tame, yet I have never been able to quite touch one. Their little feet move so rapidly that they seem to flow smoothly down the sandy beach as if they were riding on roller skates. As the waves roll in, and out again, the little drove of birds runs sideways and forward at the same time, always managing to keep just above the edge of the moving water.

It is almost a lost cause trying to search for certain kinds of small shells with them running immediately ahead. At times, to avoid a feeling of frustration, one must find the wisdom to just accept what cannot be changed, relax, and simply forget about the coveted shells while you enjoy the lovely little birds having their meal. How they do love to eat the small shells and how very constantly they go hunting.

There are more than twenty sandpiper species of varying sizes. The *Actitis macu-*

laria is the most common species in North America. About eight inches long, olive gray and white in color, it is marked with dusky spots and a conspicuous white bar on the wing. Each winter it migrates to Brazil and Uruguay to nest and raise its young.

The second beach bird companion is the sea gull (Plate 12b). Sea gulls are far tamer than sandpipers, but they dislike having their feathers touched. If you carefully respect their whims and approach them just right, you can often hand-feed them, and unless you get overanxious and make a wrong move, they will soon be fighting for a place to perch upon your shoulder.

As you go down the beach, they are really a joy, and much company, as they fly about overhead. They are very attractive, and their soft feathers are excellent for stuffing pillows, and their eggs and young squabs are a valuable food. They are very noisy birds and one becomes so accustomed to their noise that something suddenly seems wrong if they stop screaming. They love to sit on the water and to ride the waves as they look for food. They especially favor fish and living mollusks, but they are also excellent scavengers and very helpful in cleaning up refuse from shores and harbors.

Eating small mollusks presents them with no challenge. But their method of dealing with large-sized ones is one of nature's miracles. They drop large shells from the air and crack them on a hard surface, as if they were eggs. In some areas they caused such noise, clatter, and disturbance, dropping shells on buildings with roofs resembling sand or cement, that measures had to be taken to stop them. After some experimenting, it was discovered all that was needed was a little paint—to paint an ocean scene atop a roof.

In recent years more and more sea gulls fly inland to spend the winter in and around large cities where they are attracted by garbage dumps. Scientists are now making a careful study of their changing behavior pattern.

The gulls travel in groups at all times. But, as nesting time approaches, they congregate in vast numbers. It is estimated that a million or more nest in one area. They fly several miles inland and lay two to four eggs in nests built on the ground in bare rocky areas, as well as among tall grass on swampy shores.

Adapting to less favorable nesting conditions in the far north, the flexible gull manages to build its nest in small pockets and depressions in perpendicular rocky ledges along the coast. Under these seemingly impossible conditions, the gull multiplies into fabulous numbers if left undisturbed by man. Visiting a colony of ledge-nesting gulls is a memorable experience.

At Bonaventure Island, off the north coast of the Gaspé Peninsula in Quebec, the Canadian government has made such a sight-seeing experience possible. The island has been made a bird sanctuary and, for the benefit of tourists, the government operates excursion boat tours circling the island. From the boat, one has a good view of the 250-foot-high cliff with its multitude of ledges populated with an amazing number of nesting birds. The island is large enough that an hour or more is needed to circle its high-cliff perimeter, all of which is heavily loaded with not only sea gulls, but with seventeen other varieties of wildfowl nesting peaceably together.

If the rulers and citizens of our warring nations would but study the toleration and understanding of birds, each for the other, we, too, could surely find our way to better united harmony as we must live together on our various ledges of the world.

Sea Heart, or Sea Bean

We cannot leave the beach without first having a look at the lovely little sea hearts, or beans, that bring such a thrill to anyone lucky enough to find one (Plate 12c). They take a beautiful high polish, making lovely bracelets and necklaces, and are held in high esteem as good-luck charms. There is a superstition that if a person is wearing one, he will never drown because the seed has crossed the ocean through a mystery wrought by God.

They are found so very far from where they grow (often the width of an ocean away) that quite naturally their origin remains a great mystery in many areas. Finding one on the beach always brings a joyous thrill to the heart of any collector.

Actually, they are seedpods from vining tropical plants. They are carried by rivers to the open ocean. From there they are quite at the mercy of ocean currents and storm winds with their destination purely a matter of chance.

I have been fortunate in finding a few different species. There is usually only one bean species in a whole area. From my experience, it would seem the large heart-shaped bean, or sea heart, from the plant *Entada gigas* is the most plentiful. I found one beach a little south of Veracruz, Mexico, literally lined with them. This same species I found scattered along the southern Texas coast, especially in the area of Padre Island. I was told they often ride the Gulf Stream and storm waves until they eventually find their way to the coast of Ireland.

Sea beans are sometimes used as emergency food. However, they must first be thoroughly soaked and cooked for they contain a toxic material that is very poisonous.

The Lives of Seashells

WHAT IS A MOLLUSK?

9. The Mollusk and How It Grows

The Mollusk

To me, a seashell is one of the least understandable—and most marvelous—of all God's creations. And to the shell, I would presume that the thousands of collectors (scrambling over the beaches of the world in their odd costumes) are even less understandable.

To most of those souvenir hunters, a shell is merely a hard inanimate object of intriguing shape. They pick it up just as they would pick up a pretty little stone without even wondering where it came from. In fact, many believe it *is* a stone. When I ask an audience what they think a shell is, the most standard reply I receive is that "it is a kind of beach stone that an animal sometimes crawls into for shelter."

Actually a seashell is the hard outer covering, or skeleton, of the soft-bodied boneless animal—called a mollusk—living within. It would be no more possible for the shell to grow without that living animal inside than for a human skeleton to be formed without the surrounding living tissue (Plate 14 a and b). What a revelation it would be to those souvenir hunters if, just once, they could see the complete colorful animal still alive in the "stone" they so admired; admired in spite of the fact that its bleached dull color hardly bore even a faint resemblance to its former brilliance—to the color and sheen it possessed before the animal inside died and its shell was washed ashore to be quickly ruined by sun, wind, and sand.

Every seashell is a mollusk skeleton, but not every variety of mollusk carries a shell. For example, squid, slugs, and octopus are shell-less mollusks. This situation has given rise to considerable confusion in nomenclature. The term *shell* is very often used to refer to the actual animal, not just to the shell itself, to the point where the words *shell* and *mollusk* are often used almost interchangeably. But actually the shell is nothing but the hard outer covering of the *shell-bearing species* of mollusk. To avoid confusion, we shall maintain this distinction throughout; the word *shell* will be used to mean just that, and when we wish to refer to the entire animal, we shall use the term *mollusk*.

Not every kind of mollusk lives in the sea. There are freshwater mollusks (for example, river clams), and mollusks that live in trees and on land (slugs and snails). In fact, in the animal kingdom, only insects exceed mollusks in numbers. No mountain seems too high, no tree too tall, and no ocean too deep for a mollusk to thrive. Combined, the saltwater, freshwater tree and land species of mollusk cover the whole earth. They have been found in the depths of the ocean where the water pressure amounts to tons per square inch. They have been found high above the snow line in the Himalayas and in the tops of our tallest trees. They survive the tortures of heat and drought of the desert, and they even live in the steaming hot water of thermal springs as well as in the solid ice of frozen lakes and ponds.

No one knows how many different species exist, but the number is estimated to be well over sixty thousand. By far the most of these (nearly 90 percent) are very small, ranging from microscopic up to one-eighth or one-half inch. The remaining species (approximately 10 percent) vary in size, with the largest giant clam (*Tridacna gigas*) measuring four feet or more at its longest point, and weighing up to five hundred pounds.

The majority of mollusks fall into two categories. Those with two shells hinged together, like a clam, are called bivalves, *bi* indicating two shells, or valves, as they are sometimes called. Those with one shell, like the common garden snail, are called univalves, *uni* indicating one shell, or valve (Plate 14 c and d).

Most univalves (or snail type) move by extending a broad muscular foot, and then move forward on it with a gliding motion. Their migration habits have been studied by marking specimens just as bird migration is studied. Awkward and slow as their motion is, some marked specimens have been recovered up to sixty miles from the starting point of the experiment.

Most bivalves (or clam types) are equipped to either swim, burrow into mud and sand, or else attach themselves permanently to solid objects.

The mollusks with no outer shell covering, such as the squid and octopus, have a brain, whereas the shelled mollusks are just equipped with a sensory system. In most species the blood is colorless, although in some it is a pale pink, and there are a very few mollusks known to have red blood. Many species have gills, like a fish, with which they can remove the oxygen from water. Many are equipped with siphon tubes to draw water into the gills. Some are carnivorous (meat eaters) and some are herbivorous (plant eaters). Just as in the world of land animals, the smaller ones are eaten by the larger, and many are good for human consumption. Also, just as with land animals, fossil shells have been found from the Paleozoic era more than 300 million years ago.

Reproduction

A mollusk begins its life, like all complex animal species, with the union of an unfertilized egg and a male spermatozoon. For land and tree snails the problems of reproduction are simple, as they have stationary debris and soil in which to bury their eggs. But for marine mollusks, living in the waters of the ever-churning and changing ocean, the problems are monumental. The ways with which the marine mollusks cope with, and conquer, such handicaps are many and varied, and for that reason are of exceptional interest. Some give birth to live babies. Some discharge thousands of ova into the sea to be fertilized later. Others lay an egg case containing literally thousands of embryos (Plate 15).

Many mollusks are dioecious (that is, individuals are either male or female) like higher animals. Many more are hermaphroditic (with both male and female sex organs located in the same animal). In still other types, each individual of the species goes through a stage in which functional sperm is produced when the animal is quite young; then later, after that male phase, about half the population turns female to produce self-fertilized eggs while the other half remains male.

Some of the mollusk species copulate just as higher animals, but in many species, the females shed their unfertilized eggs directly into the sea while, at the same time, the males shed their sperm. The eggs and sperm float about in the water until, by accident, they come together and fertilization takes place. So many millions are released that this apparently risky meeting actually takes place in fabulous numbers. The water over the oyster beds is actually cloudy during reproduction cycles.

After fertilization in the water, the egg at once begins a process of cell division. Within a few days it is transformed into a free-swimming larval form. A special girdle of cells provided with microscopic beating cilia (hairs) encircle the larva, acting as organs of locomotion. As the larva grows, these bands of oarlike hairs extend above the edges of broad lobes that form an organ called the velum. In the next stage the hard shell begins to form on the soft body of the larva. As the larva becomes heavier, the velum is absorbed and the young mollusk drops to the bottom to continue its growth as a recognizable member of its species.

By instinct marine mollusks employ many interesting ways to protect their developing eggs. Some species place them in a skin pouch on the back of their head where the babies can simply move away from their mother when they are ready to feed. The mother moon snail secretes a gelatinous mass, puts the eggs inside, stiffens the mass with sand, and flattens it into a thin collar by smoothing it into a thin layer over the top of her shell.

Other species place the eggs into a jellylike mound for protection, and most leave them in the water. Some attach them just inside the edge of their shell. Some females glue their egg capsules to the backs of the males of the species.

Many string their capsules on a long cord.

Very few species feel any mother instinct to continue guarding their eggs after they are laid and properly protected, although the octopus and the female cowrie have both been observed guarding their egg masses.

The mortality rate of marine life is extremely high. Only a few of the millions survive to reach adulthood, while the remainder serve as food for the many other animals of the sea.

Shell Growth

Some baby mollusks begin life at birth or hatching time with their shells already formed, whereas others, as already explained, go through a free-swimming larval stage, and spend their first days unprotected by a shell. The building of the shell is one of nature's truly remarkable achievements and one that is very difficult to understand. It is one of the areas where I ran into much difference of opinion. Weighing all the information I have been able to find, it would seem the mollusk builds his shell somewhat in the following manner:

Each mollusk is equipped with an organ called a mantle, which somewhat resembles a soft fold of skin. Cells within this mantle fold secrete the material from which the protective shell is formed. The bivalve (clamlike) shells have two mantle folds—one to build each of its two shells, or valves. Each shell mantle contains a network of microscopic tubes. Each tube exudes tiny particles of lime that constantly adhere to the portion of the shell already formed. The color tubes in the mantle are arranged something like the holes on a player piano roll, allowing the color pigments to tint the shell in the right spots in the growing design. In some species some of the lime tubes rest from time to time to produce the ridges and grooves characteristic of the species. In some species some of the lime tubes are extra active and produce spines. During a shell's life, often one or more of the tubes are injured or destroyed. The remaining tubes then continue to function and produce shells of freakish, distorted shapes.

The shell structure consists of three layers. The outer skinlike protective layer is called the periostracum. The hard middle layer is called the prismatic layer because the elements of which it is composed are shaped like little prisms. The inner layer is called the laminated layer. It is very smooth and shiny, often beautifully colored, and may or may not be pearly.

The top and middle layers are secreted by the edge of the mantle, whereas the laminated, or inner, layer is secreted by the whole outside of the mantle covering. In some species, such as olives and cowries, the mantle extends to completely envelop the shell. Some authorities believe it builds new layers all over the outside surface, instead of merely extending the edges. Other authorities believe the mantle only covers the shell as a protection while building from the edge only. No cleaning problem is involved preparing these shells for display since the entire outer layer is always completely new.

In some species the animal has the amazing ability to enlarge the inside cavity by dissolving the inside shell layer and redepositing the lime on the outside surface, thus making a larger home to accommodate its growing body. If the shell grows large enough, a beach hunter can put it to his ear (that is, if the shell is empty) and listen to the "roaring of the sea." This actually has nothing to do with the sea. The same sounds can be heard by pressing other similarly shaped objects to the ear. The sounds already exist in the air, particularly those of low frequency, and are amplified by the shell, which acts as a sounding board.

Water temperature and the presence (or absence) of certain dissolved salts all greatly affect the size, color, and sculpture of the shell being built by the mollusk. There is little or no growth during seasons of adverse water temperature. The irregular periods of growth often cause growth rings in the shell.

* * *

These odd brainless creatures so amazingly adapted to every area on earth provide an ever-growing fascination for the shell collector. The larger his collection becomes, the more torture and danger he is willing to endure as he searches for just one more treasure to make it more complete. From the beginning of time, no collection was ever complete, so the addict spends a lifetime clambering from mountaintop to the depths of the sea on his joyful, never-ending quest.

10. Biological Classification of the Mollusk

The process of determining the general classification into which every mollusk species must fit has been difficult from the beginning for the malacologist (or conchologist, as he was formerly called). This process has been one of constant change, accompanied by much strife, disagreement, and often ill will.

In every generation one group of workers is regarded with apprehension by the other group, which sometimes refers to them as species mongers and splitters. The compliment (?) is returned by the others, who call the opposition lumpers.

The task of classification is a truly difficult one as well as a very important one to the world of science. But for the purpose of this book very little use has been made of it, since the general reader feels little concern with it and since the technical information is already available in scientific shell books for those who are concerned more deeply.

However, in the course of my book I do use the terms *family, genus, genera,* and *species,* so a brief explanation of these terms is in order. The family is a large all-inclusive group of *similar* animals. Then its members are sorted out into smaller groups of still closer resemblance. These smaller groups are called genera in the plural, and genus in the singular. In each genus there are still many types that differ in some slight way from all the rest, so they are separated one by one, and each is called a species, with none other in the whole world exactly like it.

So each individual mollusk type, or animal or plant of any kind, is a species. And now comes the term that I find most confusing of all to the average person. What is the difference between species and specimens? If you have one *Busycon carica* shell you say you have one specimen of the *Busycon carica* shell species. And, if you have a hundred *Busycon carica* shells, you say you have a hundred specimens of the *Busycon carica* species. And so it is with every other form of life.

In the course of my travels I ran into many longtime collectors who were disturbed because they were still unable to get a clear understanding of classification from the scientific books they had bought. To help that group of readers, I labored many hours with those difficult (to me also) scientific presentations. For their study I have outlined on the next few pages as clearly as possible the complete classification method for the whole animal kingdom down to the lowest subspecies.

Those not interested in the details of these pages may skim through them lightly, or skip them entirely since they are not vital for an understanding of what follows.

To scientists, or collectors, thoroughly familiar with the subject, my explanation of the subspecies may seem incomplete. But I ask them to bear with me as they read and remember that the purpose of this is not to serve science, but to serve the nonscientific general public and to serve those who are in trouble as they struggle unsuccessfully for an overall understanding of classification.

Kingdoms: All things fall into three basic classifications:

Vegetable Kingdom
Animal Kingdom
Mineral Kingdom

Phyla: Each kingdom is split into major divisions known as phyla (a Greek word meaning tribe).

Each phylum consists of a group of animals that are alike in some structural characteristic. For example, all animals with backbones are classified as one phylum, and all with external skeletons and jointed legs are classified in another, and so on.

It is the phylum Mollusca with which we are concerned here, or the group of creatures having boneless, soft, unsegmented bodies, which may or may not be protected by a limy external shell.

Classes: The next classification group below phyla is classes. The phylum Mollusca falls naturally into six classes:

Class I. Monoplacophora
Class II. Amphineura—chitons
Class III. Scaphopoda—tooth shells
Class IV. Pelecypoda—clams
Class V. Gastropoda—snails
Class VI. Cephalopoda—squids, etc.

Until recently the first class (Monoplacophora, also called gastroverms) was known only from fossils. It is a rare, deep-sea, primitive group. Deep-sea explorations have now brought several living species to light. The shell is cap-shaped, and it has gills, but no eyes or head. Class II chitons are small boat-shaped shells that cling to rocks and eat their algae cover. Tooth shells, Class III, are the small shells shaped like an elephant's tusk that natives often use strung in a necklace. Everyone, of course, is familiar with the general shape of clams, snails, and squid.

The class names above are more understandable when broken down into their Greek parts as follows:

Class I. Monoplacophora
Mono (one) *placos* (plate) *phora* (bearing)

Class II. Amphineura (chitons)
Amphi (on both sides)
neuron (nerve)
Class III. Scaphopoda (tooth shells)
Scapho (boat) *pous* (foot)
Class IV. Pelecypoda (clams)
Pelecy (hatchet) *pous* (foot)
Class V. Gastropoda (snails)
Gaster (belly) *pous* (foot)
Class VI. Cephalopoda (squid, octopus, etc.)
Cephalon (head) *pous* (foot)

Orders: The next step in classification is to separate each class into orders. We choose one mollusk and place it in one of the four orders, depending upon its major physical characteristics.

Order I. Aspidobranchia—limpet types
Order II. Pectinibranchia—most marine snails
Order III. Opisthobranchia—slug types
Order IV. Pulmonata—lung bearers

Families: The next step is to separate each order into family classifications to get a still closer resemblance of physical characteristics. The mollusks group into an amazing number of families, such as the four listed below:

Cypraeidae—cowries
Pectinidae—pectens
Conidae—cones
Muricidae—murexes

Genera: We then proceed to break down each family into still more finely graded classifications called genera (Plate 16a). From a large family group we make many smaller groups with a still closer resemblance.

For example, in the *Melongenidae* family we have four shells listed below. They still have important differences, but they are so very similar we place them in the same genera:

I. *Busycon*
II. *Melongena*

Species: We then proceed to break down each genus into still more closely related groupings called species.

We will now choose just one of the two mollusk genera listed in our sample above: No. I. (*Busycon*) and break it down into four of its species.

I. *Busycon pyrum* Dillwyn
II. *Busycon carica* Gmelin
III. *Busycon canaliculatum* Linné
IV. *Busycon contrarium* Linné

The four names above are names of single mollusks, not classified groups of mollusks.

Each individual species is given two names. The *first* name, always capitalized, is the *genus* name. (*Busycon* genus in this case.) The *second* name, never capitalized, is the *species* name. To these is added a *third* name, always capitalized, which is the name of the person who first named the shell species. That person is called the author of the name of the shell. He will always (with *words*) modestly disclaim any pride or elation he feels when a shell carries his name, but, as he utters those words of denial, the sparkle in his eyes belies every one of them. And why shouldn't he feel proud of his achievement? No reason at all.

Subspecies: There is still one more classification left to explain. Occasionally a shell is found that clearly belongs to a certain species. Yet it will carry some minor different characteristic that seems to justify a still finer classification.

For those special shells the species are broken into *subspecies.*

We will now choose just one of the four *Busycon* species listed above. The second one (*Busycon carica*) has been broken down still further. Two more shells of that species were found that differed enough that they were given the *subspecies* names:

First—*Busycon carica carica*

Second—*Busycon carica eliceans*

In each case the shell carries the regular species name. But it is given a third name to place it in the subspecies classification.

As early as 1000 B.C., shells had great commercial importance to the ancient Phoenician dye commerce, but no one had ever felt the need to establish a system of shell naming and classification that would be understood by all.

The earliest research in conchology was done by Aristotle. He wrote a detailed account on mollusks in 336 B.C., but still no naming system had been needed or developed.

About four hundred years later, in A.D. 77, Pliny the Elder wrote a thirty-seven-volume encyclopedia of natural history including a study on mollusks, but still no naming system had been established. Pliny was the leader of the Natural History Society of Pompeii and was killed in A.D. 79 by the eruption of Mount Vesuvius. Seventeen centuries later a fabulous collection of ancient world shells was unearthed from the ashes of Mount Vesuvius, and it was presumed to have been Pliny's collection that he had discussed in his book.

It was not until the early fifteenth century, when European traders began sailing to various areas of the world for exploration, that the need for shell naming finally arrived. They returned with odd shell souvenirs from different parts of the world, and the various merchants began competing with each other as they established their own private shell museums.

The desire for mutual discussions followed but, without names, no one could make himself clearly understood as he boasted of his new shell finds to his rival collectors. Obviously books giving names were needed—but how, was the question.

The first books were extremely crude and without a system of any kind. By the middle of the sixteenth century beautiful large books with fine hand-colored engravings were written giving names and descriptions. But still no naming system was devised. Each author, in his own language, just used his own favorite nickname for the shells in his own book.

It was not until 1758 that the first real headway was made. The great Swedish naturalist, Carl von Linné (also known as Carolus Linnaeus) felt a system should be worked out whereby a certain name for a certain animal would mean the same thing to all people of the world. It was he who perfected the system of giving each animal, including mollusks, two Latin names—a genus and a species name.

Human nature was at that time (just as it still is today) skeptical and reluctant to accept change. So the world was slow in accepting Linné's obviously sensible plan.

However, although the system has now been accepted worldwide, by no means has all controversy ended. The application of the system to individual shells has been a matter of constant strife and disagreement throughout the years of its use. Many names are still constantly changing as arguments are won—or lost—until it is impossible for the ordinary collector to keep his identifications updated and in line with everchanging name decisions.

A shell identification book that is accepted today as correct and the very latest is, five years from now, outdated in so many areas that it can no longer be used with confidence. This constant renaming and reclassifying creates a monumental problem in some cases, for which I can personally vouch.

In 1948 I began my project of collecting shells and ocean life for the enlightenment of landlocked people—people who lived over a thousand miles from any beach and were fascinated by objects from a world totally outside their experience. Realizing the importance of proper scientific identifications, I bought the latest books from recognized authorities to use as identification guides in cataloging my many finds.

Twenty years later—in 1968—so many name changes had been made I realized years of effort were all but lost since so many of my names were no longer in use. I wanted very much for the world to benefit to the greatest possible degree from my collecting effort. So I bought all new books from the younger authors who had won those many name-changing arguments with the last generation. Then I used the year 1969 to make some twelve thousand name checks to bring my catalog up to date.

It is heartbreaking to realize that the next generation of young scientists will perhaps deem so many of the present names to be incorrect that, by 1988, history will repeat itself, necessitating another year of updating effort to keep pace.

11. Something About Some Ocean Shell Families

My presentation of specific shells on the following pages branches so far from the norm that it needs a bit of explaining.

My objective in writing this section of my ocean book is totally different from the usual objective of the many shell books that already exist. Other shell books describe each single species to help the collector identify the shell he found so that he can place it in the correct shell family with its correct set of close relatives, and so that he can tag it with a correct personal, or species, name. My concern, however, is not so much to help the collector identify or classify his shell, as it is to help him learn interesting things about some of his shells.

With a twenty-five-year effort, searching the equivalent of more than seven times around the world, I managed to have a look at, or learn about, a few of nature's marvelous processes, which I am presenting here for others to share and enjoy. And believe me, interesting nature facts about mollusks are extremely difficult to come by. They perform all their life functions beneath the waters, mud, and sand, where observation is difficult. So as I traveled I carried my trusty notebook with me, jotting down a tidbit of information here, a tidbit there, and a tidbit over yonder. All the information in this section is drawn from the notebook I carried on that long journey.

I am discussing below only the families or shells about which I somehow managed to learn something interesting, forgetting about the rest. And since help with identification and classification (help that is readily available in dozens of scientific books) is not my purpose, the shells are listed in alphabetical order by their popular names for easier reference. For still further reference help, you will find them all listed in the index at the back of the book, along with many of the different colloquial names by which they are known in different areas, as well as by their Latin names.

It is my hope that you will just sit down comfortably with my book, relax, drop all deep scientific matters from your mind, and enjoy with me, and marvel with me, at nature's fantastic plan of life under the sea.

Abalone, or Ear Shell

The beautiful abalone (of the *Haliotis* genus) are some of the most handsome of the world's shells. Some California species are the largest of any shells found on the western coast of the United States, and they are better known than any other single shell of that area (Plate 16b). The shallow bowl shape with a beautiful iridescent center makes it a favorite souvenir for all tourists. The iridescent pearls occasionally formed by the abalone are also highly treasured.

Abalones off the coast of California occasionally produce very fine green pearls that are of extreme value. Prima donna Madame Lillian Nordica had a magnificent one weighing 175 grains. Its worth is considered beyond price.

California is especially noted for its green, red, pink, and black species. Many other species are widely distributed in the areas of Africa, Japan, and Australia. An especially beautiful blue green species is native to New Zealand and is called, locally, paua, and is referred to commercially as marine opal. It is so highly esteemed that laws were passed to prevent its being taken from the country. While I was there, it was only through the help of wonderful friends that I was able to add some of their special paua treasures to my collection.

There is a worldwide demand for abalone, not just for the shell as such, but because it adapts so well to the making of beautifully colored ornaments, jewelry, buttons, knife handles, card cases, and so on. Great quantities of abalone are exported to Paris and to various European countries for use in fine inlay cabinetwork, and large quantities of the green species are ground into a powder from which beautiful iridescent gems are made. The delicious flavor of abalone meat contributes greatly to its broad popularity. The broad foot muscle, when prepared with proper skill, becomes a steak feast fit for a king. It is also ground into delicious soups and chowders. The meat is especially favored in China and Japan where great quantities are consumed.

To be edible, the abalone steak must be prepared with a skilled technique. Improperly prepared, it becomes as tough as old leather.

Procedure: Place the meat inside a cloth and pound well with a stick to ensure tenderness. Roll in flour and fry or grill *no more* than 3 minutes. Properly pounded and not overcooked, it is tender as steak. It has a distinctive flavor more like mutton than shellfish. In fact, in colonial times, the name for abalone was muttonfish.

In life, the bowl-shaped shell is borne on the back of the animal. Since the animal spends its life crawling about on rocks, no underneath shell is needed for protection. The scar in the center of the empty shell is the point at which the animal's body had been attached. The live mollusk thrusts its head from beneath the shell in the manner of a turtle. It has two eyes on short stalks, a pair of long tentacles, and a broad snout. Around the outside edge of the shell is the fleshy, fringed mantle. Protruding through the row of holes around one side are long feelers projecting from the mantle underneath. Waste matter from the animal and the water that has bathed its gills are ejected through that row of holes in the shell. In very young shells there are no holes. Gradually open ones are added as the shell grows. As new ones form, older ones are later closed by deposits from within, always leaving a certain number open, the number varying with different species.

The abalone is a vegetarian and scrapes its algae and sea-plant food from the rocks upon which it lives. It is equipped with a long rasping tongue, or radula, designed perfectly for the job.

A boat must be used to collect abalones since they prefer a ten- to ninety-foot water depth. Many abalone fishermen dive for them. Others use an especially designed long-handled tool. Great care must be taken to dislodge them with the first attempt, since only the first try has any chance of success. When disturbed, they instantly clamp to the rock with a suction so strong that they cannot be pried loose without breaking the shell.

When captured, the abalone shell is covered with a rough horny coat, very closely resembling the rock upon which it rests. With proper acids and grinding, that rough outer shell can be ground away. In some species the shell immediately underneath takes a beautiful high polish. But it is the inside lining that is in great demand. To build the lining, the glands of the mantle secrete a special nacre that is pearly, iridescent, and lustrous. It is this glowing lining that gives rise to the nicknames Aurora and Rainbow shells. The commercial demand for abalone-shell linings is so great that laws have been passed in many areas regulating the size, number, and time they can be taken.

The many uses for the abalone place it in grave danger of extinction. For example, medical research has discovered that abalone juice seems to make experimental animals resistant to polio. So the possibility exists that there may be an even greater demand for these shells in the future. But it also faces danger from natural enemies. The sea otter, a mortal enemy of the abalone, was no serious threat in its rare natural state. But the otter has now been protected from hunters by law and is presently multiplying at a vast rate. It has already destroyed most of the red abalone off the Monterey coast and is feasting its way southward. The State Department of Fish and Game is now alerted and struggling for a proper solution.

One promising attempt to replenish the abalone supply is already under way. Until recently very little was known about the reproduction of abalones. But at long last scientists have finally been successful in breeding some in captivity. They learned the female lays about two million eggs, squirting them through the holes in the side of her shell, after which they are then fertilized by the male.

Four-week-old babies the size of watermelon seeds are now being flown to Turtle Bay, Baja California, where they are being fed in special temperature-controlled tanks. Later scientists plan to plant them on rock ledges at a depth of about fifty feet, in the hope of reestablishing colonies in new areas. Since it takes five or six years for an abalone to mature, much anxious waiting must now be endured before results of the experiment can be known.

Angel's Wing and Relatives

Angel's Wing

The beautiful white angel's wing (formerly *Barnea costata,* now *Cyrtopleura costata* Linné) is the most popular member of the Pholadidae family (Plate 16c). Because of its distinct ribbed formation, it is also called the ribbed pholas.

This shell rates high in popularity for many reasons. Its meat is eaten in many areas, though it is not especially favored in the United States. The lovely fragile shells, shaped exactly like angels' wings, adapt well to imaginative shell artwork.

They grow in colonies and the shell grows to approximately eight inches in length. The angel's wing is one species of mollusk where the occupant is larger than its house and cannot fully retract within it (Plate 16d). The animal has a peculiar quality, when viewed in the dark, of radiating a shining greenish glow.

In infancy it has a brief free-swimming period. It then settles down into sandy mud and bores into the home where it is to spend the remainder of its life. It lives in an upright position about a foot beneath the surface, with its large siphons extended to connect with the water above, food and oxygen needs being supplied from water drawn down through the siphon.

The chief enemies of angels' wings are crabs and man. The former nips off the end of the siphon before it can be retracted. And man does even worse. Tempted by that end of exposed siphon, you grab it firmly with one hand, get down on all fours, and start furiously digging—and digging—and digging. All the while water fills the hole almost faster than you can dig it out again—and again. Be sure and maintain your sense of humor when you go angel winging, for you are sure to look exactly like a frantic dog digging out a groundhog.

It fell within my personal experience to see the total destruction of an angel-wing bed by a severe storm. The poor things were left scattered everywhere, both above and below high-tide line. Far the majority were broken and mangled, but even the live ones were helpless to do anything but just die. From the general chaos I was able to salvage quite a nice addition for my collection.

Teredo Shipworm

The shipworm (*Teredo navalis* Linné) belongs to the Teredinidae family, which is closely related to the boring angel's wing Pholadidae family. There are many known species of this family, but the teredo is most widely known because of the great damage it has done.

These wormlike mollusks bore into submerged wood and force the United States navy to meet a huge annual bill for the repair of ships, docks, pilings, and so on. The navy is now sponsoring an intensive research program hoping to understand the life cycle enough to determine a cure. Some wood-boring parasites and barnacles will die quickly if surrounded by fresh water, but not the teredo. It has the ability to hermetically seal the opening of its burrow after the siphons are retracted. Under controlled conditions it has been found to live as long as six weeks thus sealed away from its food and oxygen supply.

The teredo was the scourge of ancient Roman galleys and of all ancient explorers. It riddled the Dutch dikes, threatening all of Holland. In 1502 Columbus, on his fourth and last voyage to the West Indies, met with disaster through the teredo's ravages to his ships' hulls. It became necessary to beach his last two vessels at Jamaica to prevent them from sinking.

In 1917 the teredo invaded San Francisco harbor. Ferry ships, wharves, warehouses, and piers collapsed and loaded freight cars fell into the sea, all because of the undermining borings of this wormlike clam. Studies proved that an increase of salinity in the bay waters, caused by droughts and shrinking rivers in the area, was the reason for this unusual invasion of the marine mollusks. The Office of Naval Research still grants research money toward the solution of this serious problem. The American Telephone and Telegraph Company also spends large sums trying to prevent the teredo from destroying its underwater, lead-encased cables, and the logging industries along the British Columbia coast suffer losses, due to its ravages, in the millions of dollars.

In spite of the tremendous problem the teredo causes, it is also a welcome scavenger. It rids the harbors and shipping lanes of floating and sunken timbers that would otherwise be a great problem and burden.

The life history of the teredo shipworm is a most unusual one. The sperm from the male reaches the female egg through the inhalant siphon tube as she draws in water. She carries the young in her brood pouch through the developing larval stage. They are then expelled into the sea as tiny bivalves for a brief free-swimming period. During this time the infant borer has eyes that disappear in later life. After they grow to about the size of a pinhead they settle on the surface of submerged timbers, attach themselves with byssus cords, and commence to bore. The tiny hole made to enter the wood is never enlarged, although the burrow later widens as it lengthens. No man-made steel bit can bore more effectively than this odd little mollusk of the sea.

The teredo is a most unusual mollusk, belonging to the same class as cockles, mussels, and oysters. But as the tiny infant bivalve matures, it loses the characteristic appearance of a shell and takes on the appearance of a worm. A fully matured teredo resembles a limp white worm up to nine or ten inches in length, with the still tiny bivalve (clamlike) shell attached at one end. It is actually those two little hard shell valves that are the cause of all the burrowing destruction. They are so constructed as to rock on two pivots on their inner surface. This movement causes the rasplike ridges on the outer surface to grind the wood into sawdust as they burrow deeper and deeper. This wood dust is swallowed by the teredo, and it is believed it uses it as food.

In time, the infected timber becomes so riddled that the borers have to turn and twist their burrows in all directions to avoid penetrating a neighboring burrow. Finally, the timber becomes so completely riddled that a crosscut section resembles honeycomb. And, amazing as it may seem, none of the trails ever puncture the wall of another. Each trail is lined with a pearly coat of lime.

Only the slender tips of two tubes, or siphons, are ever extended through the tiny pinhead-size entrance to the burrow. One siphon is used to draw in water supplied with dissolved oxygen and tiny food particles essential to the living animal. The other expels waste, mainly modified wood. When you pick up a piece of wet inhabited wood from the water, you can see the fringe of the colorless siphons quickly retract into the openings. They are equipped with small valves to quickly seal the small entrance holes.

I have been fortunate in finding samples for my collection showing different species of wood, rock, and clay borers at work in their different types of material. They add nothing to the beauty of the collection, but are of considerable educational value.

Ark Shells

Ark (*Arca*) shells of the Arcidae family are very common in all the warm seas of the world. Because of their concave shape they make ideal scoops and are the principal tools used by children all over the world to build their castles of sand on the beach.

The ark family lives from tide line to a depth of three hundred feet. They attach themselves to rocks by a glandular secretion that quickly hardens into a strong byssus or cable. However, differing from most attached shells, they can quickly cast off the cable and then just as quickly secrete another. They scramble about in rock crevices grubbing for food and then reattach themselves for another rest period.

Mollusks of this family demonstrate a keen sensitivity to light, as a protective device against predatory sea life. When a shadow passes or approaches them, they quickly close their valves in self-defense.

There are many different ark species. Perhaps the one most commonly recognized is the turkey wing (*Arca zebra* Swain.), also sometimes called Noah's ark (Plate 17a). It is very plentiful, especially on the west Florida coast, and being marked just like a turkey wing, it is in demand at shell novelty factories.

Although it is a plentiful little shell, another ark, the *Anadara ovalis* (Brug.), is considered one of the world's most unusual mollusk species because it contains red blood. Normally mollusk blood is either pink or colorless.

Band Shells

In this section we discuss several species of the family Fasciolariidae, or, as they are commonly known, band shells. It is intriguing to consider why they are called band shells, for, with fine illogic, many of these shells do not have bands. Be that as it may, they are an interesting group comprising fourteen living species, among which are included some of America's finest shells, including the largest shell found in North America.

Horse Conch

Our well-known Florida horse conch, now called *Pleuroploca gigantea* Kiener, but formerly called *Fasciolaria gigantea* Kiener, reaches a length of thirty inches and is the largest known shell of the Western Hemisphere. Since it is the largest of the band shells, in some areas it is called the giant band shell. On the other side of the world in the warm Australian waters is a shell of similar shape, but of an altogether different species (*Syrinx aruanus*, formerly *Megalatractus*), which grows slightly larger. With the exception of the huge several-hundred-pound clam of the South Seas, these are believed to be two of the three largest shell-bearing mollusks in the world.

The giant band, or horse conch, has a large brick red animal inside the huge shell (Plate 17b). It lays a large egg case in which hundreds of infants develop in the capsules. When hatched, the infants have a brick red shell the color of the meat of the parent. As they grow older, their shells usually take on the same yellowish tan of the parents' shells, although occasionally an albino develops. They are ravenous meat eaters and go crawling over the ocean floor, overpowering any other mollusk they can find. They travel over a wide range from shallow to very deep water. It is not unusual to find quite a large one while wading at low tide in the waters surrounding Florida. The horse conch is a shell specific to Florida; its habitats extend only a short distance into adjoining states. On June 18, 1968, Florida's Governor Claude Kirk signed a bill making the horse conch the official shell of Florida. He used orange-colored ink to sign the bill, and later presented the pen to the Palm Beach County Shell Club in honor of their decisive effort in getting the bill passed.

Good fossils of this large shell have been found that date back twelve million years. So it is now known that the big horse conch has been thriving and multiplying for a long time.

Tulips

The second noted band shell of the *Fasciolaria* genus is the lovely Florida tulip shell (*Fasciolaria tulipa* Linné), commonly called the true tulip. It is one of the most handsome of our American native shells.

The tulip prefers calm, shallow, algae-infested, pebbly coasts and thrives well from North Carolina southward to the West Indies. Finding one never ceases to bring a thrill. It is beautifully patterned with color and sometimes reaches an interesting ten-inch size. It makes a determined effort to escape from your hand, striking out viciously with its sharp-edged operculum, or door.

A smaller, but far more plentiful tulip, *Fasciolaria hunteria* Perry, is also found in the same area (Plate 17c). It, too, is a nicely patterned shell, and is commonly referred to as the banded tulip. It grows to only two or three inches in size and feeds by thrusting its long proboscis into the limy tubes of worm shells and sucking out the soft parts.

Both the large and small tulip species lay ruffled egg cases, attaching them to some foreign object. In my collection is an especially good egg-case specimen of the large species. It is still firmly attached to an empty glass fruit jar just as I picked it up from the water.

Boat Shells

The little boat shells, sometimes called slipper shells, belong to the family *Calyptraeidae,* genus *Crepidula* (Plate 17d). The family also includes the cup-and-saucer shells.

The most common of the little boat shells (*Crepidula fornicata* Linné) is dearly loved by children, with its appealing broad bottom, shapely stern, and little boat seat. They are useful in so many ways in furnishing seaside playhouses, and they also make perfect dishes for the young hostess to use at her doll parties.

Under the trade name Quarter Decks they are sold commercially by the ton, running into a business grossing thousands of dollars each year. Fishermen dredge up the accumulation of dead boat shells and sell them to oyster growers to line the oyster beds for the young oysters to settle upon. Great care must be used to sort out all live boats as they become a serious pest in the oyster beds. They multiply and often stack up three or four deep on an oyster to stunt its growth. They rob the oysters of food and choke them besides. No way has been found to combat the live ones once they infest the beds.

Boat shells cling to any solid surface. It is believed that most of them stay in one spot all their lives. Their shape adapts to the form of their chosen location, so no two are ever found shaped exactly alike. They frequently take on the contour of the host to which they are attached, making it very difficult to see them. The head has a long snout. The eyes are at the base of the tentacles. Some species lay their eggs underneath the foot as in a brood pouch. And all eat whatever algae and minute seaweed comes their way.

The cup-and-saucer (*Crucibulum*) is closely related to the boat shell and follows the same life pattern. But, instead of being boat-shaped with the little seat across one end, it looks like a saucer with a little cup formation in the center. Some species grow up to three inches in diameter. The most beautiful one I ever owned I had dug from a rock in the South Pacific. It was later stolen from an exhibit table where I had spread out a few unprotected shells for a small exhibit.

Brachiopods

The brachiopod stands alone in the world of shells, since there are a few bones in its otherwise soft body that keep it from being classified as a true mollusk. However, its lovely external shell (some a beautiful red and others snow white) graces the shell cabinets of collectors all over the world. Brachiopods are more commonly called lamp shells because of their strong resemblance to the ancient lamps dug from the ruins of Pompeii and other Roman cities (Plate 18a). The little pedicel, or cord, with which they attach themselves, completes the lamp illusion since it resembles the lamp's wick.

They have the honor of being one of the oldest known animal groups, and are of great importance to the geologist in dating rock strata. So, while the collector dredges into the depths of the ocean for his lovely cabinet specimens, the geologist goes searching through deserts and mountains for the ocean beds of antiquity. There he finds the many fossil remains to use in trying to solve the mysteries of creation.

In past ages there were at least three thousand different species of which a good fossil record has been made. It is believed that, at that time, they played the same important role in the world of invertebrates as that played today by oysters and clams. However, little by little, down through the ages, more and more species become extinct, until now only about two hundred living species are known.

Bubble Shells

The extremely fragile, globular bubble (*Bulla*) shells belong to the family Bullidae (Plate 18b). In life they are amazingly beautiful and are highly treasured by many people in coastal areas who use them in home, seawater aquariums. The animal, as well as the shell, is beautifully colored and provides a truly enchanting sight as it moves through the water. It is far too large to withdraw into its shell so, at low tide, it must secrete great quantities of mucus to keep its flesh moist.

It burrows in soft mud, swallowing whole small bivalves and snails, which are later crushed between calcareous plates on the strong walls of a gizzardlike organ. The thin shell is so extremely fragile it cannot be boiled and cleaned as other shells. Tiny black crabs that live on the beach rocks seem to be the best help in cleaning the bubble shell. They do a perfect, and quick, job.

There are many different species of bubble shells scattered over the world. But, like almost everything else in nature, it is the least attractive that are the most plentiful. Florida's drab little blue gray bubbles are sold by the gallon. I found them so plentiful on one beach that I could not take a step without crushing some. But in the equatorial area, it is quite another matter. Here, some are extremely beautiful, but very scarce. Some tiny species, no larger than a big pea—sometimes referred to as jewels of the sea for their exquisite coloring—will sell for many dollars each.

Carrier Shells

The carrier shells of the family Xenophoridae are the comic masqueraders of the shell world. The genus name *Xenophora* means carriers of strangers, and there are but a few species in the family. Most inhabit the warm waters of the South Pacific and Indian oceans, although we do have three species occurring from North Carolina to Cuba.

These carrier shells have an odd habit of cementing pebbles or broken shells to the top of their shell, so that from above they resemble small piles of debris (Plate 18c). So perfect is their camouflage job that, as they search for food among drifts of old shells, it is quite impossible to distinguish them until they move.

They are top-shaped, but considerably flattened. Most have a three- or four-inch width at the base, although one larger deepwater species is found in the Caribbean and in the Gulf of Mexico. They move about with a most ungainly gait. The foot is divided. The larger half reaches forward from beneath the shell and fastens to some stable object ahead. Then a mighty contraction of the foot muscle jerks the odd creature forward as the smaller part of the foot drags along behind. Thus, as it goes jumping along, carrying its load of old shells, it is truly a comic sight to behold.

It begins to camouflage when very young, at first cementing only tiny shell fragments to its tiny shell. Then, as it grows larger and larger, it uses larger and larger pieces for its cement job. And strangely, it will often use the same species of shells for its complete covering. Some species choose to cover themselves with small stones instead of with shells and they are called mineralogists. The species covered with shells are called conchologists. I was fortunate in finding good specimens of both, but my real pride and joy is a freak. It cemented just one long oyster shell that stands on end at a cocky angle like a single feather on a hat.

To do the cement job it leans against the shell of its choice at exactly the right angle, and then secretes the bonding fluid, which quickly hardens. Fossil specimens have been found to show that this shell's curious habit of self-ornamentation has persisted for countless centuries.

Chank Shell: The Sacred Chank of India

The sacred chank shell (*Xancus pyrum*), so sacred to the Hindu, is the national emblem of the kingdom of Travancore. The images of the Hindu god Vishnu always carry a chank shell in one hand. Hindu legend tells of the time the sacred scriptures were stolen by the strong chank shell. Vishnu then became a fish, chased the chank, recovered the sacred books, and destroyed the thieving mollusk.

According to legend, Vishnu had the power to change himself into any form he chose, so the Hindu readily believed the scriptural story. At the beginning of every prayer, the Hindu worships the sacred shell to ensure his prayer being heard. And each warrior carries a charm cut from the chank.

Hindu artists carve ornamental designs on the shells, and then they are used as oil vessels to illuminate the temples. Medicine is administered to the sick from them, and ceremonial oil is stored in them. Many are used for the manufacture of bracelets, armlets, and spangles. A rude saw, operated by both hands and feet, cuts the shell into rings or segments of circles. These are polished, painted, and inlaid with precious metals and gems. The Hindu women wear them in great numbers on arms and ankles, and after death each woman is buried with her ornaments. Many years ago the sepoy soldiers in the East India area wore a necklace made by stringing the "buttons" cut from the top of the shell. The button top is formed into beads, called krantha for stringing.

The shell is pear-shaped and only about five or six inches long. It has a thick solid construction, with a greater specific gravity than any other known shell. Lifting one for the first time is a revelation. Instead of feeling like a shell in your hand, it feels more like a quantity of lead. Normally the chank shell has its aperture, or opening, on the right-hand side. Very rarely a reversed, or left-handed, one is found, and it is worth its weight in gold. It is especially revered and is employed only on very special occasions.

Chitons

The chiton (pronounced kit'n), which has been on earth 500 million years, is called the coat-of-mail shell because its overlapping shells call to mind the traditional coat-of-mail garment. This odd little creature is actually a mollusk, but because of its faint resemblance to other shells, few people think of it as a seashell. Its little boat-shaped shell is composed of eight plates, overlapping like the shingles on a roof (Plate 18d). There are never more and never less than eight plates unless an injury causes a deviation to produce a freak number. Some chiton species are believed to live twenty years or longer.

Chitons are distributed worldwide on rocky coasts from the Equator to the Poles, and in the ocean down to a depth of thirteen thousand feet. They range in size from microscopic to ten inches, and come trimmed in an amazing number of beautiful color patterns. Our own northwest coast bears the distinction of producing the world's most beautiful chitons. One large species, *Cryptochiton stelleri*, found from Alaska to California, is the world's largest, sometimes reaching ten inches in length. However, the greatest number of species are found in New Zealand and Australia.

By day, the odd little creature sits in a crevice in a rock, tightly attached with its suction foot. By night, it moves about in search of algae and other vegetation, which it scrapes from rocks with its replaceable radular teeth. Often it leaves its rock entirely. It congregates in bunches around decaying seaweed. Then it returns on its slowly moving foot to exactly the same crevice on exactly the same rock to rest another day.

The chitons' rocky diet created a need for special teeth, which the all-wise Creator provided in abundance. According to one authority, Dr. Lowenstam, they have seventy iron-coated teeth in two rows of thirty-five each, each attached to the radula. Only four to six teeth appear at the mouth opening at one time. With a kind of conveyor-belt arrangement, new ones are brought forward to replace the first six that quickly become worn and drop out. At the time of each forward shift, new replacements start growth at the back of the line. Thus, that number seventy remains constant no matter how fast the front groups fall away.

Great dexterity must be used in collecting them. When disturbed, they clamp down on the rock so tenaciously with their muscles that a chisel would be needed to pry them loose. They can be easily taken by slipping a thin-bladed knife beneath them very quickly, being very careful not to announce your presence first by touching them.

When flipped from the rock, they immediately curl up like a pill bug, but will soon flatten out for cleaning if placed in a pan of salt water. After cleaning, they will recurl, so they must be bound tightly to a narrow board until thoroughly dry.

The backs of many species of chiton are studded with hundreds of minute specks. By microscopic examination, it has been determined that each of these specks is an eye—complete with retina, pigment, and lens, and connected by nerves passing through the shell. It is believed, however, that they do not actually function as eyes; they are thought to be merely light-sensitive cells.

Little is known about the reproduction of chitons, but it has been determined that the method differs in different species. In most species the eggs and sperm are shed into the water, while in some species they develop while held in place beneath the mantle. An interesting study was made of one common species by the Great Barrier Reef Investigation in Queensland, Australia. The chiton chose a night of full moon when the sea was calm and the weather warm for spawning. However, if the conditions were unfavorable at full moon, it would wait until the next full moon, ignoring all favorable weather conditions between full moons.

The chiton has the unique ability to produce mineral matter such as that produced by geological forces, and, for that reason, are an object of interest to both geologists and conchologists.

According to an article in a recent issue of *Science* (a publication of the American Association for the Advancement of Science) written by Dr. Lowenstam, Professor of Paleoecology at California Institute of Technology, the chiton can produce four different minerals without high temperatures and high pressures that are needed to do the same thing deep in the earth.

They produce two iron compounds, one of which had never before been found to have a biological source. The others produced are a phosphate and a carbonate. The first three are incorporated into its extremely hard teeth, and the other used for its protective coat-of-mail shell.

Clams

The term *clam* is used, and misused, for so many different mollusk species that the word has come to be almost synonymous with the term *bivalve*—that is, any mollusk with a two-part shell connected by a hinge. Actually, scientifically speaking, there really is no such thing as a clam—a strange statement to anyone who has ever dipped crackers into a bowl of clam chowder or smacked his lips over a beach clambake. The fact, however, is that *clam* is a shorthand term, or nickname, used to designate some fifteen thousand mollusk species from seventy or so wholly different families, all of which share only the one characteristic of a two-part hinged shell. An entire book would be needed to discuss them all.

Few of these species are beautiful, but in this chapter we will take a look at several of the especially attractive and interesting groups, as well as a few of the food mollusks that we popularly think of as clams.

Our discussion will range from the tiny pinhead amethyst gem clam, up to the five-hundred-pound giant "man-eating" clam. We will take a look at the way clams manage to survive under apparently impossible living conditions. For the most part they prefer beach areas to deep water. Some species burrow down as far as three feet beneath the sand and mud. They are nourished through long snorkel-like tubes, called siphons, which they extend up to the water above, and draw down food and oxygen. Other species live snugly inside hard rocks, within which they eat and breathe; and still other species attach themselves firmly to some submerged solid surface, and just sit waiting for their food to come to them.

As we start down the beach in search of clams, be careful and watch your step: you are going to slosh in mud over your ankles, slide over slippery rocks, and wade in water up to your neck. You may even begin to feel you are being attacked when, as you cross a mud flat, jet streams of water start to shoot up at you. But be reassured—these are just burrowing clams withdrawing their snorkels from the surface, as they sense the danger of your presence.

Amethyst Gem Clam

The tiny purple amethyst gem clam (*Gemma gemma* Totten), found on the United States Atlantic coast, is a dainty colorful shell no larger than three-tenths of an inch across. Shellcraft factories make many fancy, intricate novelties from it, but other than that I know of no function it serves except as food for beach birds and other ocean life.

Bear's-Paw Clam

The beautiful bear's-paw clam (*Hippopus hippopus* Linné) from the eastern seas is one of the most beautiful and highly prized of all the bivalves (Plate 18e). Found in great abundance, it is shipped all over the world to dealers in shells and shell novelties. Its graceful curved ridges, scalloped edges, and ivory white center make it extremely popular when made into jewel boxes. Single valves of the larger eight- to ten-inch size are widely used as table-center containers for fruits, nuts, and flowers.

Blood Clam

The blood clam, *Anadara ovalis* (Brug.), which belongs to the Arcidae family, is a small three-inch shell common from Rhode Island to Georgia. It bears the unique distinction of having red blood, an extremely rare occurrence in the shell world. It has no special value, though children love it for sand-scooping on the beach.

Geoduck

The famous geoduck clam (*Panope generosa* Gould), named by the early Indians, is found in the Puget Sound area of Washington and south to San Diego. I am told it is the largest burrowing clam in the world, and the largest bivalve shell on the United States West Coast. The shell grows to nine inches or more, but even this breadth cannot fully contain the huge, sometimes sixteen pound, animal with his large three-foot-long siphon.

Although this big species does its boring in the mud, it belongs to the rock-boring family Hiatellidae (formerly Saxicavidae). It burrows deeply in the mud, with only the tip of the long three-foot siphon exposed to the water above. When disturbed, it throws a powerful jet of water from the long siphon and retires into its burrow. Digging it out is a major operation requiring at least two men. One man must firmly grasp the end of the siphon and keep it extended to guide the diggers down to the clam. It can be collected only at extremely low tides, which occur only about three days each month. And it is only in summer that these low tides expose the neck in daylight hours. The meat of this huge mollusk is very delicious and is greatly in demand.

Pismo Clam

The official name, or naming, of California's famous Pismo clam, a member of the Veneridae family, still seems to be under scientific dispute. One authority tells me the name is *Tivela stultorum* Mawe, while other sources call it *Cytherea crassatelloides* Conrad. Be that as it may, the Pismo clam is perhaps the prettiest of all clams. The graceful, triangular shell, which can grow to about seven inches, is painted with rays of purple and hints of red. Pismo clams are not too difficult to find. Look for a telltale depression in the sand, and then just dig down a few inches. In some places farmers run plows through the sand and turn them out like potatoes in a field.

Their flavor is especially fine, a fact for which I can personally vouch. One Christmas day found me on Pismo Beach, California, where they live in abundance. While my family ate turkey in Indiana, I ate Pismo clams on the beach. They were truly delicious, but were hardly a substitute for my family on Christmas Day—even excluding the turkey.

So great has been the commercial demand for the Pismo clam, that rigid regulatory laws had to be passed and enforced to prevent their complete destruction. In an attempt to learn how best to regulate their commercial use, biologists of the California Department of Fish and Game have spent many years and much time and money studying them. In southern California and the lower Baja California Peninsula, many thousands of the clams have been marked for study to determine their rate of growth and the age to which they live. It has now been learned that they must be permitted to grow, undisturbed, approximately twenty-five to thirty years to reach the large seven-inch, one-pound size.

Quahog

The quahog clam, which lives from Nova Scotia to Florida, is perhaps the most widely known of America's East Coast shells (Plate 19a). In fact, when the average Easterner thinks "clam," he's actually thinking "quahog." Like a number of so-called clams, the quahog is actually a venus (family Veneridae), and is known by many colloquial names, including quahog, little neck, round clam, hard shell, and, when small, cherrystone. Scientifically, some authors list it as *Mercenaria mercenaria* Linné, while others call it *Venus mercenaria.*

The quahog shell was used by the Indians as wampum money, and the animal itself is the food clam known to all. If kept cold, it lives an unusually long time after being removed from the water. This quality peculiar to it means that great quantities can be shipped alive to our mid-continental cities, where the demand for it almost equals the coastal demand.

Its shell—one of the more attractive of the clamshells—is lined inside with a partial border of purple, and grows to approximately five inches. A second species, *Mercenaria campechiensis* Gmelin, known as the southern quahog, is common in the Florida area. It is not used commercially to any great extent, and its white interior lacks the purple border from which the more valuable purple wampum was made from the northern quahog.

Razor Clam

This long slender clam of the family Solenidae so closely resembles the handle of an old-fashioned razor that naturally it has acquired the name razor clam. It is also known as the sword clam and jackknife clam. Thirty-seven known species are scattered in most parts of the world except in the cold seas. The *Ensis directus* found on our northeastern coast is perhaps the most familiar North American species.

Razor clams are highly prized for their excellent flavor and, in some areas, huge quantities are used commercially. They live in colonies buried upright in the sand connected with the water above by the long siphon characteristic of all clams.

These odd little creatures, with small dotlike eyes circling their siphons, are sensitive to light and darkness. When an intruder (or sheller) casts a shadow over them, they quickly dig deeper for protection. Or, if they feel the vibrations of approaching footsteps, the siphons quickly retract, spurting their jet streams of water into the air.

Even when lying flat on its side on top of the sand, the razor clam can flip on end and bury itself in a matter of seconds. It has a very long knife-blade-thin foot that it quickly thrusts into the sand. The end of the foot then expands into a bulb, or anchor, to pull against as the long foot muscle contracts to draw the shell downward. The thrust and pull are quickly repeated until it very soon has a two-foot protective layer of sand above it.

Sometimes at low tide the sand is washed away to partially expose a colony, and they look for all the world like a field full of tiny little fence posts. If you tiptoe with great care, facing the sun to prevent both shadow and vibrations, you can grasp them with your hand. Although you may have a good grip, it is usually impossible to lift them from the hole without a shovel. For one thing, the shell edge is sharp, and the foot begins its downward pull with such speed and strength it seems to just disappear from your hand.

I found one bed especially well exposed and, just by hand, was able to collect a real feast—or so I thought. When wham! The whole bunch simply exploded from my kettle—or so it seemed. They stretched their long feet back along the shell, then, using them like a strong spring, went sailing through the air like so many arrows. There was no more sneaking up on them when they landed in the sand. In their frantic excitement they upended and disappeared from view in about the time it took me to say "good-bye clam dinner."

They are also very active swimmers in the water. They snap their two valves together and dart about hither and yon by a kind of jet propulsion with surprising speed.

Commercial razor clam fishermen simplify their task in various ways. One common method is to pour oil or table salt over the beds, forcing the clams to surface where they can easily be taken. In some areas clam hunters (with their bare feet wrapped in rags for protection) have learned how to wade in hip-deep water and very skillfully search out the clams with their feet. Success by this method, however, is a rare art achieved only by certain tribes of natives in more remote areas.

*River Clams**

Freshwater river mussels (family Unionidae) bear no relationship to the ocean clams discussed in the earlier part of this chapter. However, they are so universally known as river clams that they may feel most at home if we permit them to associate with the sea clams here (Plate 19b).

It is a most interesting family of over a thousand known species. Every continent has its own genera, and every river drainage system its own peculiar species. In the early days of the Indians, the Mississippi River bottom was alive with thirty-two different species. In the mud and sand of the river bottom they pass through one of the most interesting reproduction cycles in nature.

Each female produces about three thousand spawn at one time, each one of which is a fully formed clam of microscopic size. She waits until a particular species of fish swims by to eject her spawn. Then the tiny clams attach themselves to the gills and fins of the fish. Each quickly grows a skinlike covering to hold it firmly to its host, where it rides about and is nourished.

Oddly, each species of clam usually chooses a certain species of fish to use as host. And it so happens that the most worthless of the fish—the garfish—acts as host to the most valuable of the clam species. Even more oddly, there is in one instance an exchange of babies between

* See Chapter 15 on Ocean Jewels for a discussion of pearls from the river clam.

fish and clam. While unknowingly taking on her tiny clam passengers, the bitterling fish thrusts her long ovipositor into the mother clam's flesh and deposits her own eggs to be nourished by the clam in her gill chambers.

About six weeks of nourishment is required by the host fish before the tiny clam has grown large enough to be seen by the naked eye. It then drops to the bottom of the river where it lives on decayed vegetable matter for five or more years. Then the reproduction cycle begins all over again.

Some clams are raised commercially by collecting the spawn and putting them on the right kind of fish for their species. But in most cases it is still not known which fish belongs with which clam species. Scientists at Columbia University succeeded in raising river clams in test tubes without the host fish, but no way has yet been found to produce them on a commercial basis.

Like most things in nature, the river clam has a struggle to exist in the face of its many enemies—the raccoon, mink, otter, bear, and man—especially man.

Different methods are used by clam diggers to lift the valuable mature shells from the river bottom. The most primitive method was to wade and search for them with bare feet protected with rags.

Then dredges were devised with scooplike tools that were lowered from boats, dragged a short distance, and raised. After sorting the shells out, the remaining debris was dumped overboard. The next development was a crowfoot bar, a tool constructed by suspending a number of short chains from a long wooden bar. At the end of each chain, several blunt-end hooks were attached in a crowfoot pattern. The tool was then slowly dragged by a boat over the river bottom, where the clams lay with their shells gaping partially open. When a clam's sensitive flesh feels a hook, it instantly clamps its shell shut as an instinctive act of self-preservation. After a little dragging, the crowfoot bar was raised with the shells still tightly clamped to the hooks.

In more recent years a new power dredge, actually a long revolving conveyor belt, has been put into operation. The lower end of the belt is drawn forward just grazing the river bottom. Tubes along the side force jets of water that thrust the clams, along with rocks and other debris, onto the belt. As the belt then moves up over the boat, the clams are picked off, and the remaining debris falls back into the water at the other end of the boat (Figure 8).

After the mussels are gathered, they are placed in large vats and boiled so the animals can be removed before the shells are placed on the market.

Besides producing its occasional valuable jewel pearl, and serving as the base for our huge United States pearl-button-making industry—both of which are discussed in Chapter 15—the freshwater mussel is ground into lime, cracked for chicken grit, crushed like stone to surface roads, and is spread in oyster beds to provide a "set" for young fry.

Rock-Dweller Clam

The giant rock-dweller clam (*Saxidomus giganteus* Desh.) thrives along the

West Coast from the Aleutian Islands southward to Monterey Bay in central California.

Frequently called the Oregon clam because it supplies the large Portland clam market, the tasty animal is also widely eaten in the Vancouver area. Its shell, however, is dingy and unattractive.

Sand, or Soft-Shell, Clam

Another famous food clam, *Mya arenaria* Linné, thrives from South Carolina northward to Greenland as well as on the shores of Great Britain. The Indians called it mananosay and taught its use to the early Pilgrims who called it the soft-shell and the sand clam. In the markets of England it is known as the sand gaper and old maid. The famous Rhode Island clam chowder is made from this widely used and very popular food shell.

It lives at depths of about one foot in gravelly mud flats of river mouths, and grows no larger than the palm of one's hand. Successful transplanting from its Atlantic home to the Pacific Coast has been accomplished, and it now thrives in great numbers in the San Francisco Bay area, living in burrows made in sediment deposits in the crevices of rocky shores.

Surf Clam (Atlantic)

The common Atlantic surf clam (*Spisula solidissima* Dillwyn), found from Nova Scotia to South Carolina, has grown in importance until it ranks next to the oyster as a source of seafood in that area. Surf clams grow in such profusion that often, after violent storms, it is estimated as many as fifty million clams can be left stranded within a ten-mile coastal area.

They are never sold in the shell as are oysters and quahogs, but are processed in complicated modern plants that extract only the muscle tissue. The product has become very popular in such delicious foods as chowders and casserole dishes.

The yellowish surf clam is one of the largest bivalves found on our East Coast. It sometimes reaches eight and one-half inches, though four or five inches is more typical, and the meat from a single clam may weigh as much as a pound. Surf clams are harvested by huge clam dredges equipped with a hydraulic jet system that blows the sand away, thus exposing the clams to the digging knife. An eight-hour dredging operation often yields as much as 225 bushels of clams.

The surf-clam industry centers along the New Jersey coast in the area of Cape May, Wildwood, and Point Pleasant. In the hope of expanding future production, the United States Bureau of Commercial Fisheries Biological Laboratory at Oxford, Maryland, is now making an extensive study of the clam's biology and distribution within a 25,000-square-mile area extending from Cape Hatteras, North Carolina, to Montauk Point, New York.

Tridacna Clam

Tridacna gigas Linné, the huge "man-eater" clam (as it is erroneously called) of the South Pacific, is the largest known

Figure 8. A soft-shell clam dredge boat in Chesapeake Bay. *Robert K. Brigham, courtesy of U.S. Fish and Wildlife Service*

shelled mollusk in the world. It belongs to the furbelowed group—that is, clams with ruffled, or furbelowed, ridges on top of their shells, of which only five other species are known. Enormous specimens have been found weighing as much as five hundred pounds, and measuring as much as four and one-half feet in length (Plate 19c and d). One mollusk can provide twenty-five pounds of edible meat. Little is known about their reproduction and age, but the largest ones are estimated to be almost a hundred years old.

I have been given conflicting information regarding size. The American Museum of Natural History in New York possesses a four-foot six-inch *Tridacna* that weighs 579½ pounds. At one time this shell was proclaimed the world's largest. However, according to the latest listing of "largest" shells in *Van Nostrand's Standard Catalog of Shells,* the largest by measurement (four feet, 6.1 inches) is in Ireland, although it is smaller by weight than the New York shell.

Their massive forms rest immobile, hinge side down, on the coral reefs of the South Pacific waters. The huge scalloped valves gape apart as much as a foot at the top, exposing a broad expanse of the animal's mantle. I can personally testify that, as one gazes down upon this clam through the clear waters of the reef, the beauty of those gaudily colored mantles defies description. They are brilliant blues, greens, violets, and yellows of every shade. No two ever seem to be exactly the same shade, with the smaller cousin, *Tridacna maxima* Röding tending to be more gaudily colored than the *gigas*.

Algae embedded in the flesh of this mollusk contribute to its high coloration. There is an interchange of oxygen and carbon dioxide between the algae and the animal that is essential to the livelihood of each. The cells of the clam eventually carry the algae into its digestive tract where it is actually absorbed as food. Thus, by this most unusual plan of nature, the animal amazingly grows a portion of its food within the tissues of its own body.

The huge shells are so remarkably hard that ax handles are often carved from them, as well as many ornaments. Single valves often serve as baptismal fonts, and on the South Pacific islands, valves frequently double as catch basins for water. Large doughnut-shaped disks, sometimes nine inches across and two or three inches thick, are carved as money, especially for the purchase of wives in the Solomon Islands.

These shells occasionally produce pearls, but since the fluid with which they line their shell is not of a pearly, or nacreous, quality, the pearls are dull and of no great value. The largest one I know about, 9½ inches long and 5½ inches in diameter, with a weight of fourteen pounds, was found in the Philippine Islands. It is called the pearl of Allah and is considered sacred by the local inhabitants because, on one side of the pearl, nature had formed a startling image of Mohammed, the prophet of Allah.

It is believed by many that the huge "man-eater" clam will attack and kill a human being. This is not the case. However, its large expanse of exposed mantle is extremely sensitive to the touch. If any part of the swimmer's body or diving equipment accidentally brushes against that sensitive mantle, the shell instinctively closes its two huge valves as an act of self-preservation. The swimmer is, of course, in serious trouble if those closing valves hold him captive, but it is through no vicious intent on the part of the mollusk.

Most authorities agree that no record exists of a human ever having been killed by the *Tridacna gigas*. However, in November 1939, a personal experience account by Wilburn Dowell Cobb was published in the *Natural History* magazine of the American Museum of Natural History in New York. According to Mr. Cobb, on May 7, 1934 he was eyewitness to the death of a Philippine native who was killed by the very clam that produced the huge Allah pearl, which is still believed to be the largest in the world.

Washington Clam

The Washington clam (*Saxidomus nuttallii* Conrad) is so called because it is especially plentiful in that state, though it grows from Puget Sound to southern California. There is still much dispute over its Latin name, which may be changed.

Its seven-inch oblong shell bulges out to accommodate an extra large animal. The meat is similar in flavor to oyster and many feel it is unsurpassed in soups, chowder, and pies. The smaller surf species live at rather shallow depths in sandy areas, while this extra large seven-inch species burrows down to far greater depths in mud, and must be harvested more or less like the burrowing clams.

Cockleshells

The cockleshells, of the family Cardiidae, scarcely receive the admiration they deserve, especially the big five-inch brown cockle (*Dinocardium robustum*), commonly called the heart cockle, found on the coast of Florida (Plate 19e). Such a multitude of them inhabit the beaches, and so many are made into cheap novelties, that one ceases to think of them as anything special for a collection. Yet, in distant areas abroad, where the cockle is a foreign novelty, it is greatly admired, quite valuable, and can be used to advantage in a trade.

The cockle family is a very large one with more than one hundred different species growing in many areas of the world. It is a free-moving, globular shell with a characteristic heart shape when viewed endwise. It forms extensive shallow-water beds in the sand and mud near sea inlets and river mouths.

When walking about in sand or mud, the cockle fully extends its long foot from between its two shell valves. The tip of the foot then dilates, forming a fulcrum upon which the muscles act. By exerting a quick pull, it lifts itself with a twisting motion and flops a distance of several inches. Its movements appear extremely awkward, but it is far more agile than it seems, as you will soon know after a few of them jump from your hands or from your collecting kettle.

The most beautiful of the cockles is the *Cardium pseudolima* Linné found along the east coast of Africa (Plate 20a). It is a deep pink and grows as large as a coconut, with a capacity of a quart or more. It is a real treasure in a cabinet, and I feel very fortunate in having some extra-nice specimens in my collection.

California boasts the largest cockle (*Cardium elatum* Sowerby). It is a deep yellow color and grows to an even larger size than the pink one. This lovely shell served as the first farm machinery for early civilization. Valves mounted on sticks were used as hoes, picks, and shovels.

The *Cardium edule* Linné, found in the British Isles and the European coast, is a highly edible cockle and a very valuable source of food. It thrives in abundance and is found in all European coastal fish markets.

The colorful rose-centered cockle (*Trachycardium egmontianum*), so plentiful on the Florida coast, brings much joy to children and tourist beachcombers. Ancient beach deposits of this plentiful shell became the solid hard formation known as coquina rock. This base, along with other shells, coral, and lime, combines into a very useful building material. As early as 1512, Spaniards visiting the shores of Florida would return to Europe with a ballast of coquina rock. The historic San Marcos fortress at Saint Augustine is made largely of the rock brought from nearby Anastasia Island.

Coffee Shell

The odd little coffee shell (*Melampus coffeus* Linné), which so closely resembles the coffee bean for which it is named, lives a most unusual life in a world all its own, midway between a marine environment and a land environment (Plate 20b). It dare not be submerged, yet it must have the spray of salt water to survive. During low tide it forages for food on the beach, as any other normal marine shell. With the return of high tide, its instincts more nearly resemble those of a land animal—but not quite. It goes scurrying for higher land, but not dry land. It merely crawls up on rocks or vegetation out of reach of the tidewater, but within the range of the constantly dashing salt spray.

The little coffee shells in my collection are a constant reminder to me of the cost encroaching civilization exacts from beach life. Early in my shelling career, in 1951, I followed a narrow unimproved road southward from Florida's Everglades National Park to a little Indian settlement at Flamingo, which I presumed to be a fishing village, and where I was treated as a strange curiosity. I was told later that this village had actually been a work camp where a group of Indians had been taken for some work project. But, be the reason for their being there whatever it may, among the entire group in the little settlement, I could find no one who spoke English, so I was quite on my own in searching for accessible beaches.

Some distance west of the settlement I managed, with no road, to reach a broad mud beach at an extreme low tide. It was literally alive with thousands and thousands of the *Melampus* coffee shells. I collected until I was physically exhausted, without even making a dent in the supply.

Ten years later I had need for more,

and returned to the same beach hoping to replenish my supply. How is it possible that ten years could make such a change?

The old rutty road to Flamingo was improved. The crude little Indian settlement had been erased from the map. In its place was a luxury resort. The wide extremely productive beach had also been completely wiped from the map to make way for so-called progress. It is only one of many of my favorite shelling areas that have been completely destroyed since 1948.

Conus or Cone Shells

Cone shells, so named for their conical shape, belong to a large family of over four hundred known species. Generally speaking, they are lacking in brilliant color but, in their subdued way, they are quite beautiful and are highly favored among collectors.

They are called the poison tongue family, and can deal a fatal sting. Handling cones is a little like eating mushrooms. It is best to assume they are all poisonous until you know for sure. But, fortunately, only five species are known to inflict fatal stings—*aulicus, textile, tulipa, marmoreus,* and *geographus* (Plate 20c).

The nonpoisonous cone is equipped with a toothed radula, or tongue; in the poison species, instead of the usual toothed tongue, there is an arrangement of harpoons and venom. As it was explained to me, there is a small sac behind the head containing a dozen or more loose harpoon-shaped teeth. Close by is a second sac containing a highly poisonous venom, connected by a tube to the first sac. When an enemy threatens, or when a food animal approaches, a hollow harpoon-shaped tooth is thrust from its sac into the flesh of the victim and the poison injected. The harpoon is used most often to catch small worms and small fish for food. It is also used in defense against the mollusk-eating octopus. Every few years another case is added to the list of human victims who die from the deadly venom of one of the dangerous cones. It paralyzes the nerves to stop heart and lung action.

The South Pacific seems to be the area most favored by cones. They hide in crevices between and underneath boulders and dead coral slabs on the beaches, and are quite difficult to find. At spawning time they attach flat leaflike egg capsules to the surface of dead shells and other solid objects. They arrange them very neatly on edge, in rows, holding them in place with a tough membrane or band.

As the cone mollusk grows, it dissolves the limy substance inside its shell and redeposits it on the outer surface, thus increasing its supply of building material and, at the same time, enlarging the home for its growing body.

In life, the cone is very unattractive in its tough outer skin covering, the periostracum. With that skin removed, very pretty color patterns are revealed that are most attractive in display cabinets. The large species are in great demand in the islands of the South Pacific, where cross sections are polished and worn as armlets by local women. The smaller cones are strung into necklaces. Often a local woman will become attracted to a piece of cheap costume jewelry worn by a tourist and unknowingly trade a necklace made of rare and valuable cones, or other shells, for the gaudy trinket she covets.

The largest known cone in the world (*Conus prometheus* Hwass), which grows up to nine inches long, lives along the eastern coast of Africa. It is quite pretty with its zigzag markings of brownish yellow.

The most colorful cone, the Rhododendron (*Conus adamsoni,* formerly *Conus rhododendron*), is found in the Central Pacific. It is small, only growing two inches long, but the lovely rhododendron-colored hue of its shell makes many people consider it the most beautiful of the cones.

One of the most noted cones is the cloth-of-gold (*Conus textile* Linné) found in the Red Sea, Ceylon, and Australia areas. Named for the intricate fabric that its color pattern resembles, it grows to a four-inch length and is one of the poison species.

The most deadly cone is the *Conus geographus* found in the Indo-Pacific. It has gone on record more often than any other for killing shellers and fishermen. It is also especially noted for its ability to swallow and digest a fish larger than itself. I saw one that had been captured and preserved in the midst of such an act and it was truly a grotesque sight.

For many years the most valuable cone in the world has been the glory of the sea (*Conus gloriamaris* Hwass) (Plate 20d). It carried that honor, not so much for outstanding beauty, as for the fact that it was scarce—so scarce, that the most valuable shell find of all time was a find of *gloriamaris.* In 1838 Hugh Cuming, a noted British shell collector, was on the Philippine island of Bohol. As he turned over a rock in his search, there lay before his eyes, not one but three live, rare, and valuable *Conus gloriamaris* shells—an extraordinary event. A few years later, so the story goes, an earthquake shook that reef into the sea, and no more of the precious shells have ever been found in that area.

So rare and highly coveted was the *Conus gloriamaris* that each find down through the years was given worldwide publicity. In the next hundred years so few were found they could be counted on the fingers of one's hands. Then, in 1963, a startling thing happened. Three were found in three months' time. The resulting publicity in the New Guinea area where they were discovered finally impressed the natives with an understanding of the value of the shell, a shell that up until then they had thought of as just a shell with no more value than any other of the cones they used for food.

A number of local professional shell collectors have now been shown pictures of the shell and questioned regarding it. It is their belief that several have been eaten on the islands. In preparing their cones for food, it is their custom to break the shell away from the meat. This custom, of course, complicates the finding of positive evidence.

In recent years positive evidence of many specimens has turned up. Consequently, the value of the rare shell (a single specimen of which sold for $2,000 only a few years ago) has rapidly declined. So the most famous of all shells, which for many years was considered the most rare and valuable of all shell species, has now relinquished that honor to its successor. The choicest and most sought after cone today is the glory of India (*Conuo milnoodwardsi* Jousseaume). Only about ten specimens are known of this beautiful shell, which grows to a length of six inches.

Medical researchers are now using cone shells in most interesting experiments. The poison of the *Conus geographus* has been found to be a potent muscle relaxer, and it is hoped that it may soon be successfully developed into an anticonvulsion drug. Toxin from the *Conus magus* causes sustained contraction of muscles or spastic paralysis. Since it excites muscles, scientists are hoping to develop a drug from it that may restore motion to injured or diseased muscles.

Coquinas, or Butterfly Shells

The pretty little coquinas (genus *Donax*, family Donacidae), often called butterfly shells or wedge shells, are distributed throughout the world. Perhaps the most noted species is the small *Donax variabilis* Say, so commonly found on the beaches of Florida.

Coquinas live in droves just beneath the surface of the sand. Collecting coquinas is both a fascinating and tantalizing experience. You can be standing on what appears to be a completely barren beach. Then an extra rough wave will roll in, and out again. As if by magic, you are then surrounded by thousands of small jewel-like coquinas representing every color of the rainbow. You instantly spot special colors such as red, blue, yellow, purple, and so on, and start picking them up feeling there will be no limit to the number you can catch as they lie flat on their sides looking so helpless. But the instant you begin, they extend a small slender foot from between the valves, anchor it in the sand, and quickly draw themselves down out of sight. Often, before you have collected even a handful, you are again standing on what appears to be a completely barren beach.

The small coquinas are of special interest to the droves of little sandpiper birds that run so constantly up and down the beach in search of food. Often the birds busily probing into the sand with their beaks will reveal the location of a coquina bed.

Both the meat and the shell are of value. The meat makes a delicious soup that is considered of special value to invalids, and the shell is most adaptable in the making of jewelry, pictures, and novelties. The shells also make a very colorful and interesting addition to the collector's

Coquina Soup

Carefully wash all sand and foreign matter from your freshly caught live coquinas. Just barely cover them with water in a cooking utensil. Bring to a boil and cook until the shells pop wide open. Remove from fire and strain through a cloth. Add one-half as much hot milk as there is broth. Add a large pat of butter. Salt and pepper to taste. Serve piping hot. (Caution guests not to scrape the bottom of the bowl, as occasional remaining bits of sand settle to the bottom.)

cabinet. I have collected them from several areas of the world. Each area has its own special species differing greatly in size, color, and shape from all the others.

Cowries

The colorful glossy shell of the cowrie (Cypraeidae) family has played an important role in man's early history. Its wide use as money, and its use by local tribes as a symbol of rank, brought the family into world renown in the early centuries. Its some two hundred species are widely distributed in the warm seas of the world, but less than half a dozen are found in the cooler waters of the United States coasts.

Beautifully colored as the shells are, the animals living inside are even more beautiful. The animal engulfs the shell completely in its mantle folds, which are dotted with contrasting colors and tufted with brilliantly colored branching outgrowths (Plate 21a). The broad foot upon which it crawls is colored with equal brilliance. It would seem that the gaudy coloring would make the shell conspicuous and easily found, but actually just the opposite is true. Its bright coloring is a camouflage to blend in with the highly colored plant and animal life among which it lives.

Some of the species prefer deep water, but most thrive in shallow water. At low tide, they can be found beneath rocks and chunks of dead coral. The instant the colorful animal is touched, it retreats into the shell quite out of sight. In your hand you then have a very clean and glossy, seemingly empty, shell.

Some of the practical functions of the cowrie shell are described in Chapter 16, but there are many more. In Borneo, skin jackets are covered with overlapping rows of the small species. In India elephant harnesses are trimmed with bands of cowries. Many small cowries (*Cypraea asellus*) are used by natives in making dolls' eyes. Some island natives trim the fronts of their canoes with *Ovula ovum* (the white egg cowrie), in the belief that cowries will keep them from getting lost at sea. In India and Africa they were used in a game similar to our game of dice. A half-dozen shells were thrown, and the number landing "aperture up" counted.

Primitive people satisfy their craving for personal adornment by wearing cowries strung as beads or sewn on clothing, while civilized people satisfy the same craving in the same way by wearing necklaces, bracelets, pins, earrings, belts, and so on made from cowries.

The popularity of the money cowrie species (*Cypraea moneta*) as currency grew out of the fact that it needed only to be strung (Plate 21b). No laborious carving was required as with other shells used for money. Their discovery in ancient ruins shows their use dates far back into antiquity. Traffic in cowries made great fortunes for Dutch and English traders. They would get cargoes of them from Zanzibar and the Indian Ocean, where they had no value. Then they would sail with them to the west African coast and up the inland rivers to areas where the tribes were ready and anxious to exchange their valuable palm oil and ivory for the highly coveted money cowries. They figured largely in the slave trade. Their value as currency varied greatly in the inland regions, but on the west coast of Africa the table of currency evaluation was: 40 cowrie shells = 1 string; 2½ strings (100 cowries) = 1 English penny; 50 strings (2,000 cowries) = 1 head; 10 heads (20,000 cowries = 1 bag; and 3 heads (6,000 cowries) = 1 dollar.

A study of old trading records made by Dr. M. Schilder reveals the following amazing amount of cowrie shells exported during the last century. Zanzibar exported approximately three hundred million between 1857 and 1868. In 1858 a thousand million were taken from the Philippine Islands to be used as money, and approximately that same number were taken from the Maldive Islands in 1800. Dr. Schilder calculated that if those exported cowries could all be joined together, the chain would reach thirty-seven times around the earth at the equator.

Many cowries are given animal names because of some fancied resemblance to the animal. Some examples are the rat cowrie, serpent's head, panther, rhinoceros, lynx, stag, leopard, and tiger.

The golden cowrie (*Cypraea aurantium* Gmelin) is named for its lovely golden color (Plate 21c). The most noted representative of the cowrie family, it has considerable value, and is in great demand by collectors. It is found in a comparatively small area of the South Pacific—in the region of the Fiji, Solomon, and Loyalty islands. A live specimen can be taken only from deep and dangerous waters outside the reefs, and only a few are ever found. The scarcity of the shell, its beauty, and the dangerous collecting difficulty, all combine to make it a highly coveted shell.

The golden cowrie is worn as a symbol of rank by native chiefs. Some tribes believe that when one of the tribe dies, his spirit enters a golden cowrie. Because of this belief, few are searched for in that area. It is a favored food of a number of large fish, a fact that also contributes to its scarcity.

What a thrill it is when the long shelling trail leads to a highly coveted live-taken golden cowrie. The most unusual and the most frightening experiences of my many-year shell search occurred in the Fiji Islands where I was in search of the golden cowrie now in my exhibit. Those experiences, plus many others, are described in my forthcoming companion book.

Cymatium

The Cymatium, of which there are a number of species, lives in tropical waters from shoreline to a depth of three hundred feet (Plate 21d). It is found in sandy mud and under rocks. Due to the free-swimming larvae that drift great distances before dropping to the bottom, in some instances the same species is found in more than one ocean.

The animal itself is highly colored and is carnivorous. Some species dig a pit, hide by almost burying themselves in the bottom of the pit, and then lie in wait for their prey to drop in and be eaten.

Dentalium, or Tusk Shells

This tubular, single-valve shell (class Scaphopoda, family Dentaliidae, and genus *Dentalium*) is often called the tusk shell from its resemblance to a miniature elephant tusk (Plate 22a).

There are about 150 known species of tusk shells that vary in length from extremely tiny to approximately five inches. Perhaps the most prominent representative of the group is the polished white money tooth shell (*Dentalium pretiosum* Sby.) of the Northern Pacific, which grows to one or one and a half inches in length. An American West Coast Indian tribe used it to make their money, called Haik-wa, Hai-Qua, which was the equivalent of the wampum made from the quahog clam by the Atlantic Coast Indians.

Indian braves collected the tooth shells with fine-toothed rakes, reaching down from canoes paddled by their squaws. The shells were then sorted and strung by the squaws on fine thread made from deer sinew. To enhance their beauty, tufts of dyed goat wool and bits of colorful abalone shell were strung along with the white tusk shells. The "ligua" was the highest denomination of the dentalium currency. It consisted of a fathom-long (about the length of a man's outstretched arms) string of twenty-five of the most perfect shells and was worth fifty English pounds sterling. A fathom string of forty shells would buy a slave. Smaller shells were of lower value. The smallest imperfect ones were strung on cords of varying lengths (called kop-kop) to use in making change.

This was the only currency known by the Indians until the advent of the early white traders, who took advantage of the Indians' desire for the shell money. The Indians had learned where and how to obtain gold from the earth, but were unaware of its value. So the traders imported quantities of the inexpensive shells from the East Coast and exchanged them with the Indians for the valuable gold. The popularity of the money began its decline with the coming of the trappers of the Hudson Bay Company. To the young Indian men, blankets then became the medium of exchange. The old men still clung to the ways of their fathers and continued to hoard the shell money. But, with their death, the era of shell money came to an end on the American West Coast, except in remote Alaskan tribes where its use continued.

The odd tusk shell grows in all the oceans of the world from shallow to very deep water. It lies buried in the sand in a slanting position, with the small end extending into clear water. The mantle from the animal extends from this end to spread out and absorb oxygen, functioning similarly to the gills of a fish. At the base end of the elongated body, filling the tube is the long slender foot with a radular-equipped mouth. This foot-head extends from the large end to probe around in the sand for food. It eats minute bivalve shells and tiny single-celled animals called foraminifera.

Flamingo Tongue

The lovely little flamingo tongue (*Cyphoma gibbosum*) is found from the Carolinas southward to the West Indies (Plate 22b). It clings to the amazing sea fan animals, which bear such a striking resemblance to plants.

The long slender humpbacked shell grows to but one inch in length, and like the cowrie is entirely enveloped in its mantle. The mantle is strikingly trimmed with black, and the dainty flamingo pink shell has a very high gloss. However, a dead beach specimen soon loses both the gloss and the color.

Helmet, or Cameo, Shells

Helmet, or *Cassis,* shells are more often called cameo shells because they are most noted for their adaptability to cameo carving. Of the approximately twenty-five known species, only the thicker ones are suitable for carving, but all species are active predatory mollusks that live on the sandy bottoms of shallow warm seas, where they feed on echinoderms (Plate 22c).

The largest species (*Cassis madagascariensis*), which reaches a thirteen-inch dimension, occurs from North Carolina to the West Indies. The two smaller best-known species are the two- to three-inch baby bonnet (*Cypraecassis testiculus*), and the three- to four-inch Scotch bonnet (*Phalium granulatum*). Both are found from Cape Hatteras south to Brazil. The Scotch bonnet was recently, through legislative action, made the state shell of North Carolina.

The black helmet, also known as king helmet (*Cassis tuberosa* Linné), is most preferred for cameo cutting because of the very dark color beneath the outer layer.

Cameo cutting is one of the fine arts of antiquity. Originally, only onyx or agate composed of different-colored layers were used, the value depending entirely upon the merit of artistic workmanship. The art reached its highest perfection among the ancient Greeks and Romans. Later lava was used for a cheaper carving. Then carving was done on the *Cypraecassis* found in the Red Sea, but it was not until 1820 that shells were extensively used. The *Cassis* shell, with its three-color layers, was then found in the Mediterranean Sea in the area of Rome, and discovered to be ideally suited to the intricate carving (Plate 22d). After that, Rome became the cameo-carving center of the world. Artist fathers trained their sons, thus passing the skill on from generation to generation. Schools were established in Rome where talented people from all parts of the world went to learn the art, students paying their tuition through the sale of their completed work.

While in Rome, in 1958, I added a brooch direct from the school to my cameo shell collection. Later, with exceedingly good fortune, other additions of special interest have been made. One is a pair of shells (one male and one female), which is unusual and not often seen. Another is a carving kit showing all the tools used for cameo carving, along with different cameos in all stages of completion, including a completed one and the shell from which it was taken. The kit was added to my project through the kind generosity of a North Carolina cameo dealer who had just received it from Rome for his show window.

One especially treasured addition to my cameo shell collection is a gorgeously worked carving by a Roman artist on a large West Indian *Cassis* shell—as large as an average sheet of paper. I had not even known that these huge cameos existed, and once I did find out about them, it took ten years of searching before I actually saw one, in 1963, in Brisbane, Australia. I was told that a descendant of the original Roman artist who made them could be hired to do the carving at a cost of many hundreds of dollars —*if* the proper West Indian shell was furnished. Eighteen months later, the still continuing search came to an end. While in Florida, purely by accident, the first one I had seen in the United States crossed my path. By one of those rare streaks of good fortune that occasionally befall us all, it was presented to my project by the lady who owned it.

Janthina, or Raft Shell

The *Janthina janthina* Linné, commonly called the raft shell, violet snail, or purple sea snail, is truly one of God's unusual ocean creations. It is the most common, and the largest, of the five known species in the family Janthinidae.

These small snails, only one and one-half inches in diameter, live in great schools on the surface of the water far out at sea. They are believed to live in all the warm waters of the world, although their habitat cannot be determined since they are carried far afield by ocean currents and storm winds. Their chief enemies are sea turtles and seabirds; seabirds find them an important source of food, and so strong is their attraction for turtles that turtle fishermen search out the *Janthina* beds to do their fishing. It has no eyes, and it has no defense but to expel a little concealing cloud of violet ink.

The beautiful and extremely fragile shell is shaded a dark purple on the topside and a pale violet on the underside. The animal inside the purple raft shell is of considerable weight and, unable to swim, must build itself a raft upon which to ride the waves. It makes its raft of separate small bubbles, and, using its foot, builds it to a length of several inches. It pushes the front part of its fleshy foot along the surface of the water to trap air into cups at its tip. It then coats the bubble with mucus and presses it into place to enlarge the raft bubble by bubble. Hundreds are needed for its few-inch construction job, but it has been clocked making three or four in one minute, so the raft is completed more quickly than one would suppose. The snail inconspicuously rides its raft in an upside-down position, clinging to the underneath side. If it is accidentally dislodged, it falls to the bottom of the ocean and perishes, for it has no means of swimming.

The raft shell eats anything it can find, but the principal source of food is the small animal life floating in the sea. When a prospective meal brushes against one of its extended tentacles, the long snout thrusts forward to capture it. Well adapted for grasping, holding, and biting, the snout can hold a four- or five-inch jellyfish and quickly tear it to shreds.

The purple raft shell is both male and female—that is, hermaphroditic. But there is still much to learn about how the eggs are fertilized. One of the five species carries several thousand embryo eggs inside her body until the young are born alive. At intervals she ejects dark masses of actively swimming larvae. The other four species attach their eggs to the bottom of the raft. They are confined in capsules (several thousand in each capsule), which are attached neatly in rows—as many as three hundred to a single raft.

A specimen collected on Muriwai Beach in New Zealand was accurately studied. Its raft had seventy egg capsules attached. In each capsule there were about 500 young, thus making a total of 35,000 embryos on the one raft.

When the eggs hatch, the extremely tiny infants (complete with shell) float about near the surface, build their tiny rafts, and start feeding on the jellyfish that nature has so bountifully provided.

Very rarely is a single raft shell found on a beach. When found at all, raft shells are found by the millions. Occasionally ocean currents and storm winds deposit whole fleets or colonies of them on a beach. As yet, it has not been my good fortune to witness such an event, but I have heard the spectacle described by those who have. A beach can be quite barren one hour, and the next hour be entirely covered from waterline to high-tide line with an ankle deep mass of purple color. The color is intensified even beyond that of the shell itself because, when disaster strikes, by instinct each one empties its purple ink sac. Most in the pile will have their fragile shell broken. Those that do land alive and unbroken are quite at the mercy of the beach birds (and shell collectors). The janthina's foot is not adapted to crawling, so it cannot return to the water, although it may lie within an inch of it.

Jewel Box, or Rock Oyster

The odd little jewel box, or rock oyster (genus *Chama*), has a most self-destructive life habit. A perfectly formed one looks for all the world like a pretty, daintily colored, little jewel box—all complete with a perfectly fitting, spine-covered hinged lid (Plate 23a). But, due to its odd nature, very few perfectly formed ones are ever found.

As tiny infants, they settle to the ocean floor, sometimes beneath several hundred feet of water. They attach for life in such cramped crevices that their growth is warped and stunted for want of proper space. They are injured by other shells crawling over them, wearing down their spines, and they become overgrown with seaweed. Thus, they lose all claim to beauty. Occasionally, one grows, seemingly by accident, in a favorable location, and it is truly a thing of beauty.

While I was in Australia, a diver located a colony attached to the inside wall of a deeply submerged large pipe. Making many dives down, he laboriously chiseled them loose, one by one. They are a truly beautiful sight—lovely pastel colors with no two the same shade. Many lovely ones have also been removed from inside the hulls of our ships sunk at Pearl Harbor.

Limpets, or Tent Shells

Limpets, or tent shells, are perhaps more numerous, more varied in size, and more widely scattered over the world than any other class of shell (Plate 24a). All bowl-shaped, they vary in size from almost

One of the author's shell designs. *Robert Rickert*

Plate 14

(a), (b) *Left:* a few left-handed whelks (*Busycon contrarium* Conrad) as found on a Florida sandbar in their natural state (the largest one is a foot long). *Right:* a cleaned shell of the same species.

(c), (d) *Left:* a live *Lima scabra* Born, a rather spectacular representative of the bivalve (two-shelled), or clam, type. *Below:* a live univalve, the so-called Baby Bonnet (*Cypraecassis testiculus* Linné). Both shells are 2½ inches across.

This most unusual photograph shows the banded tulip (*Fasciolaria hunteria* Perry)—3 inches across—laying her eggs. She forms the capsules and deposits the eggs inside, and then seals the top.

Plate 16

(a) These two 3-inch shells, both found in Florida, belong to the family Turbinidae, but to two different genera. The general configuration is very similar, but detailing is different.

(b) Two abalone.

(c) Angel's wing (*Cyrtopleura costata* Linné) —5 inches long. The little piece at the top is part of the hinge connecting the two valves.

(d) A live angel's wing just after being dug out of the mud in Florida. Note that the animal is too big to retract itself into its shell. See the next picture section for a view of its siphon before it was brought to the surface.

Plate 17

(a) Turkey wing (*Arca zebra* Linné)—2¼ inches long.

(b) A live horse conch (*Pleuroploca gigantea* Kiener), with the large red animal exposed.

(c) A banded tulip (*Fasciolaria hunteria* Perry) live—taken from the Tampa Bay area in Florida. For a view of the banded tulip laying its eggs see earlier color illustration.

(d) Boat shell (*Crepidula fornicata* Linné) —1½ inches. The top picture shows the boat arrangement within the shell.

Plate 18

(a) California brachiopod (2 inches).

(b) *Top:* a red bubble (*Bullina lineata* Gray), a rare deepwater species, only about an inch long, found in Japanese waters. *Bottom:* these two black striped bubble specimens belong to the very scarce *Hydatina amplustrum* Linné species, which inhabits the waters off the Hawaiian Islands. *Robert Rickert*

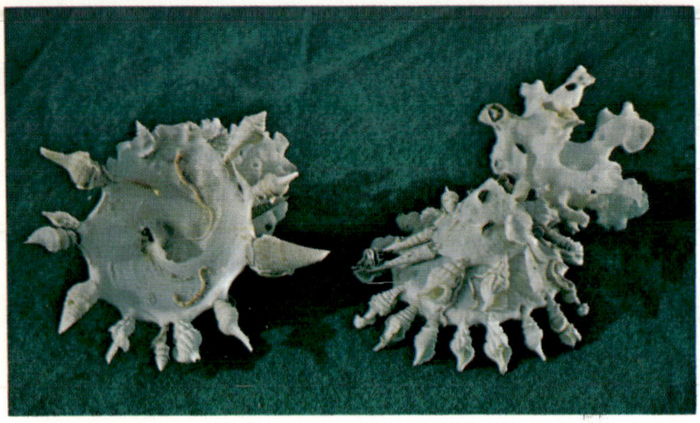

(c) Carrier shells (*Xenophora pallidula* Reeve), 4 inches across, elegantly decorated with small shells from other species.

(d) Chiton (*Chiton tuberculatus* Linné) taken in the Bahamas (2 inches long). The animal was removed from the underside of the shell, after which the shell plate was flattened to dry. The outside rim is the girdle, which holds the eight shell plates together.

(e) Bear's-paw clam (*Hippopus hippopus* Linné)—3 inches long.

(a) Southern quahog (*Mercenaria campechiensis* Gmelin)—5 inches.

(b) River clams (2½ inches long).

(c) The giant clam (*Tridacna gigas* Linné). This specimen is a comparatively small 3 feet across. When alive, the clam rests much as in this picture, hinge side down. However, the empty space between the valves was once filled with the animal's vivid mantle.

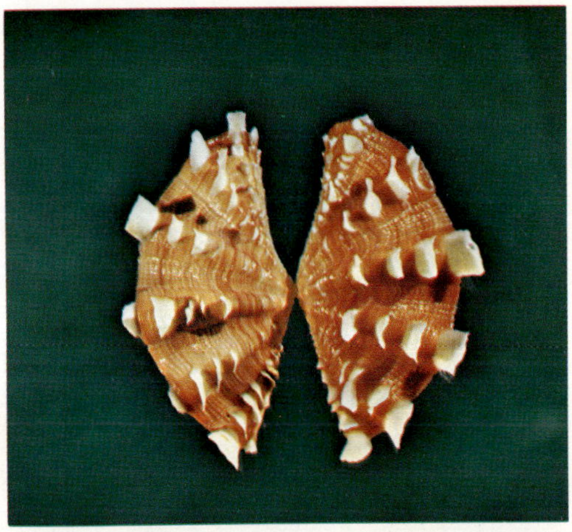

(d) Another member of the furbelowed group, *Tridacna squamosa* Lamarck. Similar in configuration to its huge giant clam cousin, this mollusk grows only from 3 to 10 inches long.

(e) The Florida heart cockle (*Dinocardium robustum vanhyningi* Clench and Smith)—4 inches across.

Plate 20

(a) African pink cockle—5 inches across.

(b) Coffee shells (*Melampus coffeus* Linné), about ½ inch long.

(c) The five poisonous cone species. Around 4 inches in length, they look like tiny delicate Chinese vases.

(d) *Conus gloriamaris* Hwass—4½ inches long.

(a) A live 6-inch-long Atlantic deer cowrie (*Cypraea cervus Linné*), the major cowrie indigenous to the United States, showing the mantle, which helps keep the shell shiny and in good repair. The mantle secretes the lime and other substances that make the shell.

(b) Money cowrie (*Cypraea moneta*)—about 1 inch long.

(c) Golden cowrie (*Cypraea aurantium* Gmelin)—4 inches long.

(d) A live *Charonia variegata* (4½ inches long) of the family Cymatiidae, taken from the Florida Keys.

Plate 22

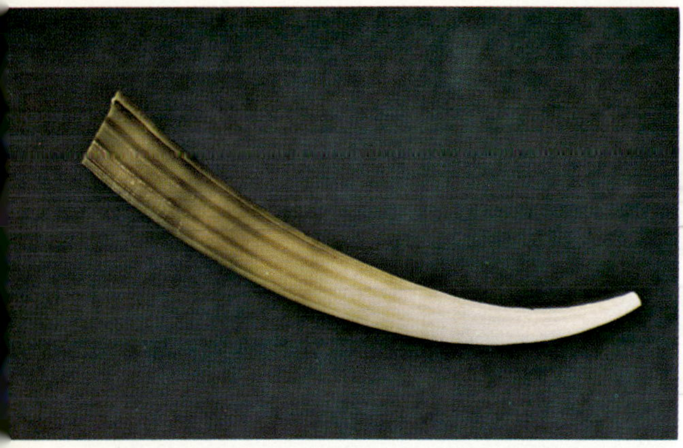

(a) Tusk shell (*Dentalium elephantinum* Linné)—3 inches long.

(b) Flamingo tongues (*Cyphoma gibbosum* Linné)—1 inch long. The shell is covered by the colorful spotted mantle of the live animal. When the animal is removed, the shell itself is plain, with no markings.

(c) A live young king helmet (*Cassis tuberosa* Linné), about 3 inches long.

(d) Five-inch-long cameo shells (*Cypraecassis rufa*), with a cameo carved into one shell.

(a) The jewel box shell (*Chama macerophylla* Gmelin)—2 inches.

(b) African land snail (*Achatina fulica*), adult (6 inches) and baby (1½ inches). *Robert Rickert*

(c) Shells (1½ inches long) of Florida's *Liguus* tree snail.

Plate 24

(a) A live fleshy limpet (*Lucapina suffusa*)—1 inch.

(b) Shell (1 inch long) of the *Littorina angulifera* Lamarck, one species of periwinkle.

(c) King's crown conch (*Melongena corona* Gmelin)—3 inches long.

microscopic to the ponderous fourteen-inch species of western Mexico—the *Patella mexicana.* This large species is frequently used as a wash basin. The small dainty species are in great demand by shell novelty factories. They make ideal little ready-made doll hats with barnacle-flower trim already attached.

In life, the bowl is, of course, filled with the body of the animal. It clings to rocks with a forty-pound suction in its foot that science has tried in vain to explain. One English species was determined to have a resistant pull of sixty-two pounds, or nineteen hundred times its own weight.

The limpet does not have the usual suction-cup action, and the foot can be cut in two, with each half still retaining the ability to grip. The muscular foot is widely eaten by South Pacific island natives who call it opihi. By painful experience, they learned that great care was needed in removing *all* the small-toothed radulas, or tongues, from the opihi. If eaten, they become entangled with each other forming a fishhooklike ball that lacerates the human stomach.

To breathe, the limpet must leave a tiny space between the shell and the rock upon which it rests. In this relaxed position it can easily be eaten by its crab and bird enemies, or flipped from the rock by the thin-bladed knife of a shell collector. But nature has provided it with an extremely fast reflex. If the first try is a failure, the limpet clamps down instantly with its forty-pound suction and is quite safe and secure from the crab, bird, and collector.

In New Zealand there is an odd little freshwater limpet (*Latia neritoides* Gray) that produces luminescence, a phenomenon that is extremely rare in the world of mollusks. Like most other luminous forms, it shuns strong light and will soon die if kept exposed to light. This limpet, which is vegetarian, has a well-developed eye and appears to be able to see light rays.

When the limpet is disturbed, a luminous mucus fills the groove between foot and mantle, which, in darkness, has a striking greenish glow, and is even visible in diffused daylight.

Littorina, or Periwinkles

The small *Littorina,* or periwinkle, is a large genus of more than 150 known species (Plate 24b). Perhaps the most widely known representative is the common little periwinkle shell (*Littorina littorea* Linné). For centuries it has thrived abundantly on the shores of northern Europe. Not until the nineteenth century was it accidentally introduced into this country. It is believed to have arrived at Nova Scotia (by way of Iceland and Newfoundland) in the cracks of wooden ships. Then later, during the past century, it migrated southward to Maryland. There are no records of the periwinkle ever having been found in American fossil beds, but shells have been discovered in excavations of early Viking settlements in Newfoundland.

Periwinkles have long been an important food in the European market where they are bought extensively by the poor and middle classes. They are roasted and sold on the streets of London as pennywinkles. Thousands of tons are collected each year on the rocky coasts of the British Isles for this street market. They are also extensively collected for fish bait. The name *periwinkle* is an old one and is believed to be a modified form of *penny winkle*, a term long used to distinguish this small shell from a large whelk (or winkle), also used as a staple food in England. During the 150 odd years since its introduction into Nova Scotia, it has spread abundantly through our New England coastal areas. However, as yet, it has never been widely accepted as a food in this country.

The shell is small, seldom more than an inch long, and is usually quite a dull color. However, in some species, the periwinkle color pattern is so beautifully designed that they are regular little art treasures. It has an unusually long radula, often three times the length of the shell, with six hundred rows of curved sharp teeth, seven teeth per crescent-shaped row. Knowing its ravenous algae-eating habits, oyster farmers often scatter quantities of periwinkles over their beds to help control the smothering algae growth.

The more northern species are equipped by nature to stand extreme cold and the pounding of severe waves. They can often be seen actively crawling about tide pools in freezing temperatures. Great numbers will live clinging to rocks, piers, and seaweed between the tide lines. The more southern species often live on aerial roots of mangrove trees, where they get only ocean spray.

They reproduce by forming eggs. Some species hatch them within the body, whereas other species attach egg masses to rocks and seaweed.

Melon, or Bailer, Shells

Melon, or bailer, shells belong to the genus *Melo* of the Volutidae* family, and are so named because they resemble a large melon in shape, and because natives use them to bail water from canoes; a large shell easily holds a gallon of water. They are a tropical shell growing in the Indian Ocean and all along Australia's Great Barrier Reef.

The *Cymbium* genus of the same family is found on the West African coast. They are so very much like melon shells that, in many collections that I have studied, all of both types are identified either by one name or by the other.

The huge animal living inside the shell has a very large foot that it drives into the sand like a shovel as it crawls. It likes to lie almost submerged in the sands and coral banks of the reef, leaving but a small center of the shell back exposed. It gives one quite a start when one stoops to pick up what one believes to be a small shell and it proves to be but a small part of a large melon shell.

The eggs are formed inside capsules arranged on a central core often nine inches long. The young reach an advanced state of development before they hatch, and are perfect miniatures of the adult. In my rather extensive melon shell exhibit I have infants protruding from the capsule, and also an extra large fifteen-and-one-half-inch adult. It seems nearly impossible to believe the one could ever grow to the size of the other.

* See page 165 for other members of the Volutidae family.

My fifteen-and-one-half-inch specimen is by far the largest melon shell I have so far seen in the United States. It is in my collection only through the kind interest and generosity of Mr. and Mrs. A. J. Read, owners of the House of 10,000 Shells, at Cairns, North Queensland, Australia. In their shell exhibit (widely known at one time as the largest in Australia) they had three fifteen-and-one-half-inch melon shells. With uncontrolled overcollecting, melon shells have long since become increasingly scarce. The Reads explained that it had been many years since any had been found of a size similar to the three old ones in their collection, and that my chance of finding one of that size was absolutely zero. Then, before I left the area, through their generous desire to share a treasure with the United States, they made a difficult decision. They rearranged their display case with but two of the precious shells, and sent the third with me back to the United States for people here to enjoy in my exhibit.

Another special treasure in my melon exhibit is a salad-bowl set made by craftsmen in the Philippine Islands from nine large melon shells. One shell was used for the large bowl, and separate shells were used for the spoon, fork, and each of the six serving dishes.

Melongena, or King's Crown Conch

The king's crown conch (*Melongena corona* Gmelin) is one of about a dozen species of the *Melongena* genus found in the warm waters of both hemispheres (Plate 24c). Being very active, it is an especially interesting shell to watch. With siphon extended, it moves forward on a large muscular foot through shallow brackish areas of mud flats. It enfolds other shells in its huge foot and smothers them. It seems almost no shell is too big for it to tackle. I am told that the large horse conch, and occasionally a large tulip, are the only shells known to overpower the active, two- to five-inch king's crown.

They are very clever hunters and, to compensate for their own small size, the king's crown hunt in groups. Often several can be seen surrounding a guardedly closed bivalve shell that they have disturbed. They just wait patiently until the smothering bivalve gapes its shell apart to breathe. Then, they quickly intrude their long tongues and devour the meat of their victim. They even work in a group to conquer a large whelk by patiently waiting until the whelk lifts its door, or operculum, to breathe. Then, quickly, they thrust their long tough snouts beneath the edge of its operculum. The victim quickly clamps them securely, but must soon relax again long enough to breathe. With each breath, the long snouts (which resemble black shoestrings) are forced in a bit farther. When they reach the muscle, the rasping tongues soon end the battle and their victory is won.

The *Melongena corona* (or king's crown) is an especially interesting species for a collection, since they differ from area to area. All in each colony will be alike. But colonies no more than twenty-five miles apart will be noticeably different in color and structure. The *Melongena* specialist can often look at a *Melongena* collection and tell within a matter of a few miles where some of the specimens were collected.

At egg-laying time, the *Melongena* attaches neat little rows of white disks to old shells. Later, several tiny shell babies hatch from each disk.

Murex

The *Murex* genus of the family Muricidae contains over 250 known species, and is one of the most lively and most widely known of all shells. The famous royal purple or Tyrian purple dyes of antiquity were made from the murex shell. The mollusk secretes a yellow fluid. Down through the centuries, the valuable dye was made from this fluid. The secret of the process was lost during the Middle Ages and then rediscovered centuries later. In more recent years the very valuable dye has lost all of its popularity to the less costly chemical dyes that have now been perfected. In ancient times the royal houses of Persia, Babylon, and Syria all wore Tyrian purple, and the major wealth of Phoenician merchants was derived from it.

These Phoenician merchants were the greatest navigators of ancient times, operating between the Orient and the more modern British world, and they held a monopoly on the highly coveted Tyrian purple. Eventually, the treasured dye was valued at something like $10,000 per pound, and the emperors of Rome limited its use to their own raiment. Thus, purple became the symbol of royalty, and the modern use of purples and crimsons in our Christian church is an outgrowth of that ancient precedent.

The first oil tanker of the Shell Oil Company was named the *Murex,* and many of their tankers and lubricants are given seashell names.

Perhaps the most widely known *Murex* species of modern times is the beautiful *Murex ramosus* Linné found in the Indian, western and central Pacific oceans (Plate 26a). The beautiful white shell with its rosy pink trim sometimes reaches a length of twelve inches and is cherished in homes all over the world as a table or mantel decoration. The attractive black murex and the pink-mouthed murex of the western Mexican coast are also popularly used as ornaments in the home.

The unusual Venus comb murex of the Indian Ocean, Japan, and North Australia is also well known and cherished by collectors for its evenly spaced long spines, so very closely resembling the teeth of a comb (Plate 26b).

Various species of the murex are found on tropical and subtropical shores of both hemispheres, from low-tide line to a several-hundred-foot depth. They are especially admired for their spinelike growths protruding from various ridges. These spines have an odd growth method. They grow by spurts, as weather and food conditions vary. At the close of each growth spurt, the ridges thicken at the edge of the shell as a kind of barricade to guard the temporarily closed doorway. Then, when the next spurt of growth takes place, these protective ridges interfere and must be partially dissolved away by an acid secretion.

The flesh of the murex is eaten by locals, but it has never been used widely as food. The animal is carnivorous, and many of the species feed on dead crabs and fish. Some feed on bivalves and cause great loss to oyster farmers.

Mussel (Saltwater)

The mussel family (Mytilidae) is a saltwater mollusk (not to be confused with freshwater clams, also called mussels). The oval-shaped shell is usually a deep purple, but is sometimes green or brown. This is a large family, abundantly represented in many temperate to cool areas of the world. A five-inch length is not unusual in the large species, but one to three inches is more nearly the average size.

Acres and acres of mussels cover mud flats and inlets in wide river mouths, where tide and currents meet. They hang in masses on piers, rocks, or driftwood near low water level, attaching themselves with byssus threads spun from a gland in the heel of the small foot. The threads are coarse and strong and are also used to spin nests from bits of shell and sand (Figure 9). Mussel beds in various areas act as barriers protecting lowlands from floods. An artificial jetty is soon loaded with mussels and filled with silt, which year by year increases its stability and efficiency as a breakwater.

The most important species of the mussel family is the *Mytilus edulis*. It is widely cultivated and a major source of seafood in western Europe and along the coasts of the North Sea. The mussel industry only began on a large scale about 1950, but it has grown rapidly until millions of pounds are now produced yearly for the food markets.

The *Mytilus edulis* is unusual in both its manner of locomotion and of reproduction. Although attached by byssal threads, it is not immobile. It has the power to drop off the old threads and spin new ones as it reaches forward with its foot. At breeding time, the mantle becomes transformed into brood chambers. The *Mytilus's* reproductive organs are so enormous—much larger, proportionately, than in any other mollusk—that the minute yellow eggs seem to fill all the space between its two shells. When the eggs hatch, the small "fry" settle upon any available support to keep from suffocating in the soft mud on the sea bottom.

Sea-bottom conditions must be favorable for commercial production. The young fry (or seeds) are machine-dredged from the shoals, shipped to mussel farms, or parks, and sown in thin layers. The parks, sea-bottom areas, are chosen for optimum tidal current, temperature, salinity, and nutrient value of water conditions. Alternatively, if no favorable sea-bottom area can be found, they are then planted on suspended ropes or on posts set in the mud with a six-foot low-tide exposure. Three years are needed for the fry to reach marketable size. During that time the farmer must constantly struggle against total destruction by starfish. Since starfish will die more quickly than shells when exposed to air, the usual method of battling them is to dredge out both shells and starfish, wait until the latter die, and then return the shells to the water.

Figure 9. A group of blue mussels found on Unalaska Island in the Aleutians.
V. B. *Scheffer, courtesy U.S. Fish and Wildlife Service*

In preparation for market, the mussels are cleansed of sand by piling them in thick layers in areas free of drifting sand. From these storeroom areas they are later taken to packing sheds where the shell clusters are broken up, washed, and shipped.

The mussel animal is also widely used as fish bait and as fertilizer. The shell often takes a very beautiful polish and enhances the collector's cabinet. It is also extensively used in shell novelty factories.

Most of the species have such thin shells that the polishing must be laboriously and painstakingly done by hand. The most beautiful work I have seen was at Sarina, North Queensland, Australia. Mrs. Myrtle Tandy, who did the polishing, was truly a wonderful person. She had a large shell collection in her home that I thoroughly enjoyed. And she sent me back to the United States with a gorgeous display of her dainty hand-polished Australian mussels for people back here to enjoy in my exhibit.

Sea otters find mussels especially to their liking as they forage along coastal shell beds for food. An otter hauls up a rock about five inches in diameter, puts it on its chest as an anvil, grasps the mussel in both paws, and bangs it on the stone until the shell breaks. Not only do the otters use the stone as a tool, but they have been observed carrying the stone under their armpit as they continue their search for the next shell, reusing the same stone over and over. According to K. R. Hall of Britain's Bristol University, the otter is the only mammal (except those of the highest order) known to use tools to obtain food.

Oliva

The Olividae family, commonly called oliva shells, are much in demand by collectors for their beautiful high gloss and attractive color patterns (Plate 26c). This tropical family of some one hundred known species is found chiefly in Central America, the South Pacific, and Philippine areas, although some species are found in many other localities.

They gather in colonies just beneath the surface of the sand where they go burrowing for food, leaving a molelike trail in their wake (Plate 26d). They seem to migrate rapidly. After one tide, literally hundreds can be found in a colony, then, after the next tide, nothing. They just disappear. The best time to look seems to be after the very lowest of morning tides. Their breathing tubes extended to the water above closely resemble the trunks of tiny burrowing elephants.

The mantle of the *Oliva* mollusk completely envelops the shell, building from the outside. Thus, the oliva shell is a joy to find, with its always clean high-gloss surface. Different species vary in size from the tiny miniatures to an approximate four-inch length. The meat is edible, although not widely used as food. The lovely glossy shell is not only coveted by collectors, but it adapts to many uses and is in great demand by shell novelty factories. In ancient times the Indians strung certain species to use as bead money, and many were sewn on garments as ornaments. In modern times they are equally popular for jewelry and many kinds of trinkets. They are especially popular for door portieres, made by filling open doorways with strings of them.

Pecten, or Scallop

The *Pecten,* or scallop shells (of the family Pectinidae), are possibly the most widely known, and certainly among the most beautiful of all shells.

No other mollusk bears the romantic and historic distinction enjoyed by the pecten. The pilgrim scallop (*Pecten jacobaeus* Linné) is common on the coast of Palestine. A member of the First Crusade, as he started home, picked up a pretty pecten shell and wore it in his hat. The custom soon became established. On down through the centuries, whoever wore the pecten badge was recognized throughout the Christian world as a crusader returning from the Holy Land.

The Order of Knighthood growing out of the Holy Wars adopted the St. James scallop (*Pecten maximus*) as its badge of distinction. Proud old families used the pecten symbol in ceremonies honoring their illustrious ancestors. So long as the Age of Chivalry is remembered, the scallop will be remembered as its emblem. Today it adorns the coat of arms of many prominent families including those of Sir Anthony Eden and Sir Winston Churchill.

The pecten was also a symbol of religion and art in ancient times as well as throughout the Middle Ages. Venus, the Goddess of Love, was believed to have been born from a scallop shell, so it is now said the scallop symbolizes resurrection, when the dead rise from their tombs just as the goddess arose from her shell. The great Venetian artists used it in their magnificently carved ornaments and in their paintings. The famous old ships of the Venetian doges were trimmed with huge gilded scallops. In the ruins of Greece, archaeologists have found statuettes of Venus stepping from her scallop shell. On the garden wall and in the ancient chapels of Pompeii, the scallop is carved lavishly. It frequently adorns coffins and monuments excavated from ancient Roman and Asia Minor burying grounds.

The Shell Oil Company chose the pecten as its emblem. Prominently displayed at every Shell gasoline station, it is, of course, recognized by travelers all over the world. The Shell Oil Company derived the emblem from a nineteenth-century predecessor company in England that had imported shells and various other goods from the Orient, and had used the scallop as its sign.

One of the deepwater pecten species of special beauty is the lion's paw (*Lyropecten nodosus* Linné), also called the knobbed scallop (Plate 27a). Their beauty, the difficulty in collecting, and the comparative scarcity of these shells all combine to make them expensive and highly coveted by collectors. They thrive among deepwater sponge beds, and are often brought up by sponge divers. Often one is found entirely embedded within a sponge growth. The sponge-producing coastal area near Tarpon Springs, Florida, is a favorite lion's paw region.

Most of the pecten species are beautifully colored; the animal is oftentimes even more brightly colored than the shell. The beautifully fringed mantle, circled by a row of some thirty eyes, is exposed to view as the shell valves gape open. Each eye is an iridescent green spot encircled by a rim of turquoise blue (Plate 27b). Studies have been made, but it seems there is still disagreement as to whether or not the mollusk can actually see objects with those tiny eyes.

The swimming action of the pecten is a most fascinating thing to watch. Each jet of water expelled by the quick contraction of the adductor muscle forces the pecten forward two or three feet, sometimes more. Each thrust forward is always in a straight line. But by controlling the direction of the jet stream, pectens run a zigzag course. Trying to catch them as they dart hither and yon is truly a fascinating sport—and what a beautiful sight they are sometimes on a bright day with the water clear and smooth. They look for all the world like a group of brightly clad youngsters having a great game of tag.

The foot of the pecten is a fingerlike organ used for spinning instead of for walking. Its byssal gland secretes a glutinous mass manipulated by the foot into a rope before it has time to harden in the water. Young scallops can attach themselves to solid objects with the byssal cord, and then sever the attachment at will to be free-swimming again. Often, during storms, they attach themselves to seaweeds to avoid being washed out to sea.

A few pecten species attach themselves permanently while quite young, and remain firmly attached the remainder of their lives. The large California species (*Hinnites giganteus* Gray) is a good example. It grows grotesquely misshapen as it conforms to the outline of the rock to which it is attached. By far the larger percentage of pectens use the byssus attachment only in their youth, and then become free-swimming for the remainder of their lives.

All mollusks have an amazing ability to repair their shells when they are broken or chipped. In my collection are a number of specimens repaired by natural healing after serious injury. It seems nearly impossible to believe the unprotected mollusk could live on to repair such extensive breakage. Pectens have been extensively used in laboratory experiments in an attempt to learn exactly how the lowly mollusk can repair its castle with such uncanny skill. It has been learned that it first coats the injured areas with a layer of soft organic animal matter called conchiolin. The mantle then places small calcium droplets over this organic matter. Next it fills in the area between these hard small drops with the regular shell material. With the foundation for its repair work thus laid down, it proceeds to add layer upon layer of the regular shell material on top until the area of repair assumes the proper original thickness.

There is one shell in my collection that had been seriously cracked, but not actually broken away. No trained human welder with the most modern of tools could possibly have done as beautiful a welding job as had been done by that brainless animal.

Over two hundred species of pectens are found scattered over the world from shallow water down to 1,200-foot depths. They have a tendency to assemble in great numbers on shallow plateaus in the sea known as scallop banks. It is to these banks that the scallop fishermen go to supply the world with the delicious seafood we call scallops.

Scallop fishing has become a truly important United States industry. The deep-sea pecten (*Placopecten magellanicus* Gmelin) thrives in many areas from the Saint Lawrence River to the Virginia capes. Large fleets go out dredging for these pectens and stay out a week before returning to shore with their catch. Huge dredge rakes are operated from the boats. They are constructed of many heavy metal ring links, and one dredger can cost as much as $200,000 to build. Often dredges are operated from both sides of a boat, thus doubling the investment. The boats are equipped with spotlights to maintain twenty-four-hour fishing.

These large pectens reach six to seven inches in diameter, and immediately after being dumped from the dredge, they are shucked right on the boat—that is, cut open to remove the edible portion commercially known as scallops. The scallop meat is packed in ice for the remainder of the week. As soon as the boat returns to port the scallops are sold at auction to buyers who have come from far and near to meet the incoming boats at the dock.

This edible portion, or scallop, is actually the adductor muscle of the pecten, or the muscle attached to the inside of the two shells or valves. It is the pulling force used by the mollusk to tightly close the two valves for protection when danger threatens. The adductor muscle is also the organ of locomotion for the pecten. By quickly closing the valves with that strong muscle, a jet of water is thrown out to force the shell forward. With its constant hard use one would assume it would be very tough, yet it is tender and succulent as can be. (Even more oddly, the adductor muscle of the permanently attached, wholly inactive, oyster is so tough it must be discarded with the shell.)

Hundreds of tons of scallops are brought in yearly to supply the rapidly growing demand brought about largely by the advent of quick-freeze processing. In past years, although popular in coastal areas, the scallop was unknown in the diets of inland people.

When it became possible to freeze scallops for long-distance shipping, the Scallop Festival was born to spread the good news. An annual festival has been held the past ten years in New Bedford, Massachusetts, which overlooks the harbor where 75 percent of the nation's scallop fishing is done. Along with the festival queen and accompanying pageantry, a bountiful feast of French-fried scallops is provided, with invitations extending to the entire country. So popular has it become that as many as thirty thousand guests representing forty states have come to consume seventeen thousand pounds of delicious fresh-caught scallops.

Scallops are usually eaten French-fried, but also can be baked, jellied, chowdered, broiled, and even "scalloped." The tenderness of the scallop, its excellent flavor, and its adaptability to many types of recipes assures the continued growth of the market.

When the edible muscle is removed,

A special New Bedford salad is made from scallops:

Cut 2 pounds cooked scallops into bite size; combine with 1 cup diced cucumber, 1 cup sliced celery, and ¼ cup sliced stuffed olives. Add ½ cup French dressing, and chill several hours. Add 1 cup mayonnaise, and salt and lemon juice to taste. Serve on watercress in the scallop shells.

A favorite scallop recipe used in the Orient is:

Line broiler pan with aluminum foil; arrange 2 pounds defrosted scallops on it. Brush generously with mixture of ¼ cup honey, ¼ cup prepared mustard, 2 teaspoons curry powder, and 1 teaspoon lemon juice. Broil slowly for 10 minutes. Turn scallops and brush with sauce again. Broil 10 minutes longer until brown.

the remainder of the animal, or the rim, is thrown away. Cats are extremely fond of eating these rims. A shell authority, Horace F. Carpenter, says that, if cats are permitted to eat these rims, their ears will drop off. Along the coast of Rhode Island where much scallop fishing is done, many rims are available, and a good many earless cats are running around in that area. Although Mr. Carpenter has made an extensive study, I find it is just one more area where scientists are not in complete agreement. So, I will just chuckle at the cats, have fun with my project, and show no partiality to either side of the argument.

The shell, as well as the meat, of the pecten is of great commercial value. The extremely colorful species of the warmer areas are exceptionally popular with collectors. Both the shallow water and deep-water species are held in high regard. The larger species, some as large as dinner plates, are popular as kitchen utensils. Many are used as individual baking dishes in which to serve scalloped oysters. Others are just used as candy and nut dishes. Millions are used by shell novelty factories, for they adapt beautifully to the making of trinkets so popular with tourists.

Pen Shells

The wedge-shaped pen shells of the Pinnidae family resemble the lower tapering point of an old-fashioned quill pen (Plate 27c). The name originates from the scientific name *Pinna* used by ancient Greeks for this shell, probably because of its shape. The name is based on the Greek word for feather or wing. The large family, ranging from four to eighteen inches in size, inhabits the warm seas of both the Old and New Worlds. They grow attached by byssus threads to rocks beneath the sand. To form the threads, a gummy secretion from the foot gland is forced through a sievelike membrane filled with fine holes. The resulting threads are usually of a brown color and have a silky texture.

The *Pinna nobilis* of the Mediterranean is famous for the extra fine quality of its golden byssus threads. It is believed to be the Golden Fleece of mythology, and in ancient times materials woven from it were of great commercial importance. It is so fine that a pair of gloves made from it can be folded inside a walnut shell. A large scarf can be put in a snuffbox and a large cape can be drawn through a finger ring. The industry persisted in a limited way in Taranto, Italy, on the Ionian Sea until after World War I. Gloves and other small items are now being woven there for curio stores to satisfy tourist demand.

The large muscle of the pen shell is an important food source along the western coast of Mexico, in Japan, on many islands of the South Pacific, and it is now finding its way into Florida menus. In Polynesia pen shells are carved to make decorative articles, and the valves of the larger species are used as plates. Pearls of moderate value, and occasionally a black one of considerable value, are produced by the pen shells.

There is an odd and intimate association between the pen shell and the tiny little crab *Pinnotheres*. The crab lives inside the mantle cavity of the shell for protection, and it, in turn, protects the shell by warning it of approaching danger.

During violent tropical hurricanes, both pen shell and tiny crab suffer mutual upheaval and destruction. The violent churning of the water reaches down to tear the shells loose from their anchors and then dumps them on the beaches by the truckload. It has been estimated that, by a single storm, more than a million pen shells have been tossed from the Gulf of Mexico onto a single twelve-mile stretch of beach on Sanibel Island off the west coast of Florida.

Most of the fragile shells land cracked and broken. And even those that survive the beach landing must perish, for they have no way of returning to the sea. Along with the shells, thousands of the tiny guest crabs must also perish. Those that survive the crash feel the impending doom awaiting their host so they crawl away from their natural protection and can be seen by the thousands as they swim about in confusion in the surf and in tide pools.

The Emperor Shell: Pleurotomaria

For many years it was believed the Pleurotomariidae family, to which the famous emperor shell belonged, was extinct. Over eleven hundred fossil species had been found, but none alive. Then in the nineteenth century five species were found—two in the West Indies, two in Japan, and one in the East Indies.

There are now known to be fifteen living species—five in Japan, one off South Africa, one in the East Indies, and seven in Florida and the West Indies (Plate 27d). They are in great demand by collectors and sell for a high price, although most specimens I have seen have been quite dull and faded in color, unless very young.

The one in my collection, which belongs to the Japanese species *Perotrochus hirasei*, is unusually bright in color for one so large. It is known as the emperor's top shell. Before 1940, no one was permitted to own an emperor's top unless he received it as a personal gift from the emperor. It was considered a great honor to receive such a gift. History tells us that at one time, during a war, the life of a man was spared only because of the emperor's top shell in his possession. A great attempt was made to keep the shells from leaving the country, so, for many years, the only ones reaching the cabinets of collectors in the United States had to be smuggled.

Polinices, or Moon Shells

The *Polinices* and *Natica* shells of the family *Naticidae* are also called natica shells, moon shells, or cat's-eye shells. This smooth apple-shaped family has many species varying in size from the miniature to five inches or more in diameter, and is widely distributed in all the temperate and warm areas of the world.

With an amazingly large foot extending three times the width of the shell, the mollusk burrows beneath the surface of the sand searching for bivalve shells to eat. The foot is also used as a gliding organ of locomotion on the surface of the sand. On top of the foot is a thick roll of flesh that folds back over the head for protection as the burrowing foot drags the animal through wet sand (Plate 28a). When a prospective shell meal is encountered, the fold rolls back from the head, a long proboscis reaches out to contact the victim, and while the wide foot holds it firmly, the tooth-edged radula unrolls to bore a round hole in the shell. Through the small hole the soft parts are then eaten by the sucking mouth of the *Polinices*. While young, they eat more than a third of their weight in clams each week.

When all soft parts of its body are extended, the *Polinices* shell is almost completely covered. The instant danger threatens, it quickly retreats into the shell and tightly closes the opening with its fragile, transparent door, or operculum.

The eggs are laid in a sticky mass of clear jellylike fluid secretion, which is then molded over the outer surface of the shell in a thin layer called a sand collar. A layer of fine sand is pressed into this collar as a protection for the eggs. So long as the collar remains in the water, the mucus secretion keeps it tough and leathery. But if it is cast ashore, it becomes dry and so brittle it can scarcely be handled. Near hatching time the sand falls away to expose the eggs to view.

Spider, or Scorpion, Shells

The many-pronged spider, or scorpion, shells belong to the Lambis genus of the family Strombidae (Plate 28b). Its ten known species are native to the warm waters of the Indo-Pacific. None have been found in American waters. The different species, ranging in size from three inches to a foot, move forward with an awkward leaping motion as they push themselves along with their muscular foot, the pronged extensions having nothing to do with locomotion as is often mistakenly believed.

My several years' struggle to learn from scientific researchers the feeding habits and reproduction method of the spider shell has been a most discouraging one. Since the whole show takes place beneath sand and water, it is easy to understand the difficulty of finding the *one* answer to each question that is *scientifically* correct, and, of course, there is but one such answer.

I have high regard for the many dedicated scientific researchers who are devoting their lives to a search for these answers. They have endured so much physical distress during fieldwork, and have spent so many weary hours bent over their desks with books, pen, and paper, that, by comparison, I feel as if I have done almost nothing at all. And they have acquired so much knowledge that, when I sit across their desks from them trying to learn, I feel very uninformed indeed, as well as *most* confused:

Confused at the different answers I get when I know there *is* but *one*.

Confused at the number of times I am told: "Oh! He [or she] is no *authority*," when the personality, background, and the desk of the "he" or "she" is equally as impressive as that of the speaker.

Confused at the *divergence* of loyalties, each for the other. In my search, both in foreign parts of the world and in widely scattered areas of this country, I find a complete lack of unity in loyalties regarding who *are* and who are *not* authories. Even in the same locality, I run into this difference of opinion.

Confused at the number of times I am told with a positive tone of conviction: "Oh! There *is* no *controversy* on that issue—the answer is thus or so"; when I have just been told elsewhere that it is something entirely different (with an equal tone of conviction).

Confused that, with book authors as well as with personalities, I have yet to find the very first one honored by all. No matter what the book, or what the cost, you cannot go far without being told: "Oh my, don't trust *that* book. The author is unreliable."

However, with so many highly educated and dedicated researchers striving to unravel the mysteries of the sea, much headway has already been made. I feel that those of us who lean so heavily upon them in our attempts to learn have every reason to believe that, before many more years pass, much less confusion may prevail.

Spondylus, or Spiny Oyster

The beautiful *Spondylus* of the family Spondylidae (sometimes called the thorny oyster or spiny oyster) was known and eaten by the early Greeks long before most shell families had even been discovered. Approximately eighty fossil species have been found, and about seventy are known to now live in various parts of the world—each different from all the rest, and each found in its own locality and nowhere else.

Of the several areas represented in my collection, to me, our own species (*Spondylus americanus*) is the most beautiful (Plate 29). In many instances I feel other areas of the world have outscored America in the beauty of their shells, but not in the Spondylidae family. It grows 150 to 200 feet beneath the warm waters of the Gulf Stream, more than a mile out from Florida's east coast.

Spondyluses grow firmly attached to rocks and other solid objects, and the divers who chisel them loose must do battle with the swift current and large sharks that frequent the area. The shell must then be very carefully processed for several weeks to remove the lime, coral, and sponge formations with which they are coated. Due to their extremely fragile nature, it is little wonder that only a few survive the collecting, transporting, and cleaning operations. So it is understandable why the few whose spines arrive at a display cabinet still beautifully intact are of considerable value. The broken inferior ones can be bought for a very nominal amount, but a really prime one is a fair-sized investment.

The long spines and wide variety of attractive colors make the Spondylidae family one of the most coveted of all for the display cabinet. We are told those long spines are built by extra-active mineral-producing glands in the mollusk's mantle. Knowing that makes the monumental task seem none the less impossible.

Strombus, or Conch

The strombs, or conchs, belong to the family Strombidae, of which about seventy species are found in many temperate and warm areas of the world. The term *conch* is applied to so many different shells in different parts of the world that it demonstrates well the scientific necessity of using Latin names for each separate species.

The name *king conch* is commonly used for the West Indies *Cassis* shell. The term *horse conch* is applied to Florida's altogether different *Pleuroploca gigantea* shell. The term *queen conch* is used for the still different *Strombus gigas* of the Gulf of Mexico. In the South Pacific the conch is the triton shell—still different from all the rest. In the New England area a conch is a still different large shell of the *Busycon* genus, formerly called the *Fulgur* genus before the heated controversy of many years duration was finally resolved and the name *Busycon* accepted by both sides.

In this chapter we will use the term *conch* as applied only to the Strombidae family—more commonly called strombs in most areas where I have been.

During the warm months of the year strombs congregate in colonies in shallow water to spawn, and their egg mass is about the most unusual in the whole world. The eggs are attached to a long jellylike tube which, when uncoiled, extends to a length of seventy feet or more. Sand grains become attached and the tube coils into a tangled spongelike mass. It is estimated to contain the amazing number of 400,000 eggs. When first hatched, the infants go through a free-swimming stage before settling to the bottom to continue the normal adult life habit for the species.

Strombs have long been considered the acrobats of the molluscan world. They have a slender, but very strong, muscular foot, with which they can do amazing things. At the end of the foot is a slender clawlike operculum, with which they close the shell opening when they retreat inside for safety. But, differing from most shells, they also use the clawlike appendage as a weapon to fight off an enemy and as an aid to locomotion. I can personally vouch for its use as a weapon since one pierced the skin of my hand and drew a bit of blood in its attempt to escape my grasp.

Each time strombs make their comic way across the beach, they demonstrate the use of that odd claw. Instead of crawling on a broad foot as most mollusks do, they extend that narrow foot far forward, jab the sharp claw into the wet sand as an anchor, then quickly contract the foot muscle to jerk the shell forward in a crazy zigzag fashion. They sometimes actually turn a complete somersault, as if they were playing. It is truly an amazing sight, watching the long leaps and quick turns of frightened strombs as they try to escape capture. With that strong foot they quickly right themselves, if placed on their back. They can even make an escape from a rowboat, and experience soon teaches one to be on constant guard lest they flop out of one's shelling kettle back into the water.

Fighting Conch

Because of its extreme activity, the little *Strombus alatus* Gmelin found so commonly on Florida beaches is usually referred to as the fighting conch. Much scientific controversy has existed for many years over the naming of this common little shell. As it was explained to me, after much strife and disagreement among those most concerned, it was finally resolved that there were two different species of very similar, and confusing, formation.

Florida's common *alatus*, usually with a smooth shoulder, occasionally grew shoulder spines. In that form it so closely resembled West India's *pugilis* as to be frequently misnamed—and thus the long-standing confusion arose.

Queen Conch

Of the many conchs in the Strombidae family, the great pink conch, or queen conch (*Strombus gigas* Linné), is perhaps best known (Plate 28c). It is one of the world's larger mollusks, growing a foot or more in length and to a five-pound weight. It was unknown to the Old World until after the discovery of America, for it is found only in the area of the Caribbean Sea and the Gulf of Mexico.

Since its discovery, great quantities have been shipped to Europe for the manufacture of porcelain, cameos, and ornaments. And it is now found in homes all over the world sitting on mantels, being used as flower vases on center tables, or made into hanging flower baskets. Its wide lip, with a layer of pink nacre over white, adapts beautifully to cameo carving. However, since the pink color is so prone to fading, it has never been used as extensively for that purpose as has the regular cameo *Cassis* shell. Occasionally pearls of exquisite beauty are produced from its iridescent, colorful nacre. They are elongated, oval in shape, and of various colors—pink, red, white, yellow, occasionally a deep rich brown, and rarely one of gold is found. They are considered to be among the most beautiful pearls in the world. The rarest and most perfect command a high price, though they too have the disappointing quality of fading. Very few conch shells form pearls, and no way has been found to produce them commercially.

The female of the pink conch is much larger than the male. She lays her egg mass in the sand beneath ten to seventy feet of water, and the young keep pretty well buried beneath the sand to escape their many enemies. Many kinds of fish, crabs, octopuses, and large mollusks feed upon them. The spotted eagle ray is a very special enemy. The ray has flattened plates in its jaws and it crushes the conchs, eats the soft parts, and spits out the shell pieces. As many as forty-one conchs have been found in a ray's stomach at the same time. Loggerhead turtles feed upon them, even killing and eating the large adult ones. Only a very small percentage of the millions hatched live to maturity.

Those that survive congregate in groups and travel over the sea floor in search of food like flocks of sheep—sometimes two hundred or more in one bunch. Each individual has two well-formed eyes at the extreme end of flexible stemlike organs, with a small elephantlike trunk for a mouth. When eating, it extends this mouth as much as four inches.

The supply of pink conchs has now been badly depleted by overcollecting, but occasionally nice ones venture into water shallow enough for them to be caught by wading at low tide. Usually they are picked up by using a glass-bottom bucket from a boat. When located in this way, they are then lifted by pronged hooks, by divers, or by dredges.

By an odd kind of cooperative arrangement, the pink conch is indispensable to the livelihood of a small dependent species of fish. When danger threatens, the small inch-long fish darts into the mantle cavity of the shell and is given perfect protection.

While I was searching a beach far south of Veracruz, Mexico, a native came ashore with a boatload. As we struggled to communicate, without aid of a single spoken word, our hands were waving and pointing in all directions. The shells were so tremendously dirty it was only a guess, but I suspected they would be extra nice ones when cleaned. At long last, the torturing, speechless transaction was completed. Then, still only on the strength of my guess that something lovely was concealed beneath the dirt, I faced the problem of shipping these heavy shells, with their fragile lips, from southern Mexico to Indiana. It was not until months later, after my return home, that both the guess and the shipping proved successful. They cleaned beautifully and are a lovely addition to my collection.

To keep their catch alive for market, fishermen punch holes in the shells, tie five conchs together, and place them in shallow water. There they tug futilely in opposite directions as they try to wander away, but they stay in place until needed.

The meat of the pink conch is eaten extensively along the coast of Mexico and in the islands of the West Indies. But for years, many Floridians, believing it to be a carrion eater, called it the buzzard mollusk and looked with contempt upon the natives, giving all the people who ate this shell the derogatory nickname "conchs." However, in more recent years, the delicious flavor of the meat has been winning reluctant Floridians over. According to the Florida State Board of Conservation, 56,356 pounds of conch meat were caught and eaten in 1966.

In the capital of Nassau alone, approximately 80,000 conchs were caught and sold each month. To help prevent extinction, the Bahama News Bureau asked the Lerner Marine Laboratory on Bimini, a branch of the American Museum of Natural History, for a preliminary survey. By 1970 extinction seemed certain unless radical measures were taken. The Florida Department of Natural Resources then passed laws levying heavy fines against any wholesale or retail fish dealer who sells the queen conch for any purpose other than food. They can now be taken by collectors only in small daily numbers in restricted areas.

It is now predicted that, eventually, vast undersea conch farms will help solve food shortages, just as the present farming of other shellfish is already doing.

So important is the pink conch to the economy of the Bahamas that its picture has decorated their postage stamps for many years, and now it ornaments their coins as well.

Tellinas, or Sunrise Shells

Of the three-hundred-odd *Tellina* species of the family Tellinidae perhaps the most noted is the lovely sunrise shell (*Tellina radiata* Linné) (Plate 30a). Its highly polished white valves, trimmed with rays of deep pink, make it a thing of beauty. Were it not for the fact that it is plentiful and easily collected, it would most certainly sell at a high price, for it is in great demand by collectors, curio stores, and novelty factories.

Many species of the *Tellina* are brilliantly colored. They are widely spread just below the surface on sandy or muddy shores of most seas, but they are especially numerous in the Indian Ocean.

Thais, or Dye Shells

Of all the shells God created, He seemed to be the most generous with the *Thais*. No matter where in the world I have searched, it seemed that although nothing else could be found, if there was a rock anywhere in sight for it to cling to, the *Thais* and its family would be waiting to greet me.

The *Thais* animal (as well as the *Murex*) secretes a fluid called purpurin, which may be green, scarlet, or deep purple, and which is often used in the making of dye. A species native to the Mediterranean was used in ancient times (along with the *Murex*) in the manufacture of the famous Tyrian purple dye used by royalty, thus the name dye shells.

One of the most plentiful of the *Thais* is the rock purple (*Thais lapillus*). It is believed to have migrated to America from Europe by way of Iceland and Newfoundland. Although it grows to no more than an inch long, some are quite pretty with a color range from yellow through orange to a deep purplish brown.

The Florida dye shell (*Thais haemastoma floridana* Conrad) grows slightly larger than two inches (Plate 30b). It is found on the shores at low tide all the way from North Carolina to Texas. Being farther south, one would expect it to be more brightly colored. But, contrary to norm, it is a duller color than the northern species.

Tonna, or Cask Shells

Tonna, or tun shells of the family Tonnidae, are more commonly called cask, or wine, shells because of their shape, which resembles ancient wine casks (Plate 30c). Many do not grow beyond the size of a doubled-up fist, but the largest eastern Australia species, called the beer barrel tun, grows much larger. Its swollen shell is capable of holding a half gallon of liquid.

Tuns inhabit the warm waters of Florida, the West Indies, the Mediterranean, and Indo-Pacific areas, and are rather attractive in their subdued tones of cream and brown. They crawl about on the sand just below low-tide line on a huge expandable foot, and their bodies seem far too large to retract into the shell, but by some miracle they do manage to squeeze inside. When they approach a bivalve or other food animal, they grasp it with the huge foot and smother it. They also attack direct by thrusting their long flexible proboscis into soft body tissue and scraping away the flesh with their sharp-toothed radula.

The flesh of the tun is edible but not considered delicious.

Paper Fig

The delicate figshell (Ficidae family) is closely related to the tun (Plate 30d). Its shell, though paper thin, can withstand the pounding of rough waves. After storms, many figs can be found strewn over the beaches in excellent condition.

Only a few species exist, scattered over the world in widely separated areas; the species from each area are similar in shape to all the rest, but different in sculpture and color pattern. Our Atlantic Coast species (*Ficus communis* Röd.) is quite plentiful from the Carolinas on south.

Figs crawl rapidly on the sandy bottom in rather deep water, carrying their shells gracefully and with ease. The gracefulness of their motions makes them a favorite in home saltwater aquariums. They thrive quite well in confinement, and provide much amusement with their long necks extending forward carrying their heads high, while their large black eyes seem to look intently in all directions.

Triton, or Trumpet Shells

Something like 150 species of triton shells (family Cymatiidae) crawl about beneath the tropical waters of the world, down to a depth of approximately three hundred feet. Some species are quite localized, while other species, whose larvae float great distances on the surface of the water, are distributed worldwide. At a certain stage in the development of these mobile groups, the larvae metamorphose into the adult form and settle down, perhaps thousands of miles from the area of the parent mollusk.

The triton shell was named after Triton, the mythological sea god of the Greeks—half man, half fish who would blow on his trumpet shell to calm stormy seas. So from ancient times down to the present, the widely used triton shell has also been known as the trumpet shell (Plate 31a).

The trumpet shell has traditionally served many and varied functions. The natives of the South Seas called the tribes to worship with the triton shell horn, much as we use church bells, and assembled tribes to battle as well. In the Mediterranean area, it was customary to make a deafening noise on a triton horn when an unpopular or ill-mated couple got married. The Minoans of Crete used triton horns in religious ceremonies. In China left-handed triton horns were especially revered. They were stored in pagodas and blown to still the waves. Once or twice a year, at high noon, Aztec priests would blow the triton trumpets, and people would bite each other's tongues and ears to present the blood to the Sun God. A somewhat similar ceremony was enacted by sun worshipers of Indonesia. The Incas had vessels decorated with pictures of people blowing their triton horns. The Maoris knocked the apex tip off and added a carved mouthpiece that produced a doleful sound. This "pu tatara" was used in times of war, as well as for other signaling. The Fiji natives use the triton trumpet to send messages over the mountains. Its deep booming sound can be heard for several miles.

Fortune smiled on me during my stay on the Fiji Islands, and I became the proud owner of a triton trumpet that had been used a lifetime by an old Fiji chief who had just died. I was told that, after his death, to keep the evil spirits away as he made his journey to the Great Beyond, the villagers took turns and kept his horn blowing continuously day and night for two weeks. A special lip technique, which I have so far been unable to master, must be used to blow the horn. My granddaughter has mastered this art and an occasional guest sometimes blows it successfully, so I know the big boom is inside, just waiting.

The tone of the trumpet horn is produced just like that of any other wind instrument. By blowing through a hole, which has to be drilled into the shell in just the right way, the air inside the shell is set in vibration. Each shell has its own basic pitch, which can be only slightly raised or lowered with practiced lip technique. But even within these limitations, the shells can be used as actual musical instruments. Perhaps the only seashell orchestra in the world is found in the little German village of Neuendettelsau, near Nuremberg. It originated some fifty years ago in faraway New Guinea.

Two German missionaries, Keysser and Zahn, sent from the Evangelic-Lutheran Church of this tiny German village, worked for twenty-five years with the New Guinea natives and the triton horns to produce recognizable hymns. Using large *Cassis cornuta* shells for the deepest bass notes, and small triton shells for the highest pitch, they were able to work out an accurate and complete tone range. So skillful did they become, they even worked out a two-voice harmony.

When Keysser retired in 1946 and returned to Neuendettelsau, he brought with him a complete set of the coordinated musical shell horns to train other missionaries. For twenty some years now the orchestra has been active in the middle of Europe, hundreds of miles from any beach.

The largest species of the Cymatiidae family are the triton trumpets (*Charonia tritonis*) of the South Pacific, and the species *Charonia variegata* Lam. of the Florida area and the West Indies, which are eighteen inches long. Pacific trumpets are the ones most highly treasured by collectors, most often seen by tourists, and most widely used by natives. Besides their use as horns, the natives frequently employ them as water containers and teakettles. The latter are fashioned by hanging them from the bottom and using a plug in the cut tip end, which serves as a spout. While in the South Pacific, I was fortunate in obtaining a beautiful large shell for my collection. I also found a young one no larger than your little finger.

This same large *Charonia tritonis* is now an important element in a multimillion dollar research project sponsored by the Queensland government of Australia. In recent years the large crown of thorns starfish (*Acanthaster planci*) multiplied excessively to upset the balance of nature in certain areas of the Great Barrier Reef, doing extensive damage to other life on the reef.

It has recently been discovered that the large triton is a natural and fatal enemy of the starfish (Figure 10). The government took measures to protect and encourage the multiplication of the triton, with the hope that it would, in turn, protect the reef against destructive starfish, but the problem eventually took care of itself. (See p. 48 for more about the starfish destruction of the Great Barrier Reef.)

Trochus, or Top Shells

Trochus, or top shells of the family Trochidae, are a shallow- to deep-water tropical shell found most plentifully in the Indian and South Pacific oceans.

With some *Trochus* species, the entire structure is made of a lustrous mother-of-pearl nacre. They take a beautiful polish after the outside color layer is ground away, and are used for making pearl buttons and ornaments just as mother-of-pearl oysters. Fifty beautiful buttons can be made from one shell, and the fragments are used for the nuclear beginning of cultured pearls (see Chapter 15). The *Trochus* thrives well in the Palau island group, about five hundred miles from Manila. They are a main source of income, and have been gathered for generations by the islanders, who must dive for the shells into thirty or forty feet of water. The reefs were regarded as clan property, much as a family regards its fields.

The great top shell (*Trochus niloticus* Linné) of Australia and the Indian Ocean is the largest and heaviest of the family (Plate 31b). The five-inch deep-water species makes lovely cabinet specimens, either polished or unpolished, and is widely used in the mother-of-pearl industry.

The magpie top shell, or West Indies top shell (*Livona pica* Linné), is the most noted top of the West Indies area. This four-inch shell takes a good polish and makes an attractive cabinet specimen. It

Figure 10. A triton eating a starfish. *Courtesy Womentco Miami Seaquarium.*

lives in great numbers among rocks near shore, and its empty shell is a favorite adopted home of the land-acclimated hermit crab. One never ceases to feel startled at the sight of a magpie top, being moved about by a hermit crab, crawling about above high-tide line or even up a tree.

The most plentiful tops found on the Florida and California coasts, usually called ringed tops, or channeled tops, because of their fine circling parallel grooves, belong to the genus *Calliostoma.* The largest grows to no more than one and one-half inches in size, but they are quite pretty and interesting to collect.

The tiger top shell (*Calliostoma tigris* Martyn) is New Zealand's favorite top. Though only two inches, it is quite pretty with its dainty, subdued color pattern. I found it quite plentiful along the west coast of New Zealand's North Island.

During low tide, the shallow-water top species can be found clinging to rocks. However, they are recognizable only as top-shaped bulges on the rock surface. Much homework must be done to remove a heavy coating of lime, barnacles, seaweed, moss, and so on, before their varied and beautiful color patterns can be seen. The tops are vegetable or algae eaters, and, in some of the species, the animal is even more beautifully colored than the shell.

Besides being used in modern pearl button and ornament factories, top shells are also used extensively by more primitive people as ornament. Colorful necklaces are made from the tiny species, while the larger species are sliced, polished, and shaped into arm and ankle bands.

Turbos

Turbo shells of the large Turbinidae family, so-called because of their turban shape, are especially noted for their unusual opercula. In most mollusks the opercula are of leatherlike pliable materials, while the turbo's doors are hard and thick, and often very beautiful. The animal can withdraw completely within its heavy shell and, with its hard thick door drawn tightly in place, it is perhaps more secure from danger than any other mollusk. Turbos feed on vegetation and are found in most oceans in areas of short weed growth near the shore. The meat, which plays an important role in the economies of many countries in different parts of the world, is considered a delicacy by many people, and is used extensively in China.

Perhaps the most widely known turbo is the beautifully colored, glossy *Turbo petholatus* Linné, commonly called the cat's-eye turbo. The elaborate blending of color on its marblelike glossy surface is most skillfully executed, and no two are alike, so it is greatly in demand all over the world as a cabinet specimen. However, beautiful as turbo shells are, their opercula (called cat's eyes) are even more beautiful. Their small, hard, glossy doors are brilliantly colored in shades of green, burnt orange, and so on, and adapt perfectly to the making of earrings and other fine jewelry. The Maori tribesmen of New Zealand used them as eyes for their god idols; they were widely employed in the temples of India, and they are now sold in almost all jewelry and shell novelty stores. Each turbo species has a different kind of operculum but, though all are unusual and lovely, none are as beautiful as the green cat's eyes.

Fifty years ago thousands of turbo opercula were sold at drugstores as eye stones, to be laid on human eyes to remove foreign matter. If placed in a saucer of vinegar, the opercula from some species are most fascinating to watch. They will move in circles as they ride currents set in motion by chemical reaction.

The turbo is also renowned for the quality of its mother-of-pearl. Though I have never read or heard of its producing a jewel pearl, its nacre, or mother-of-pearl formation, is just as beautiful as that of the pearl oyster.

The big green turbo (*Turbo marmoratus* Linné), common in the Indo-Pacific, is the largest of the turbo species (ten inches in height and in diameter), and is the one most esteemed for the quality of its mother-of-pearl (Plate 31c). The shell has a green outer layer, but the thick shell layer underneath is made of an extra fine mother-of-pearl nacre. When polished, it reflects dainty rainbow colors, and no more beautiful substance can be found anywhere in the ocean. The shell makes an especially nice cabinet specimen, either polished or unpolished. It is used for making buttons, buckles, inlays, and so on, and is especially popular as table lamps. In ancient times a Scandinavian monarch would mount the shell on silver and set it with gems for his royal drinking cup. Its lovely thick operculum, with a diameter of three to four inches, known in India as the Disk of the Moon, is frequently used as a paperweight.

Another beautiful turbo species is South Africa's *Turbo sarmaticus* Linné. It, too, takes a high polish and is a real treasure in any collection. On the rocky shores of southern Australia is a species, *Turbo stamineus*, which for centuries has been a valuable food source. Immense numbers have been found in the kitchen middens of the aborigines.

Turritella, or Screw Shells

The *Turritella*, or screw shell, family numbers over one hundred known species found chiefly in the Old World, with very few native to America. Gregarious in nature, these shells live in large colonies and crawl about shallow muddy sea bottoms in search of bits of seaweed and decaying matter.

The great screw shell (*Turritella terebra* Linné) is the shell after which our common metal screw is patterned (Plate 30e). The idea came to the philosopher Archimedes as he observed the action of the slender five-inch pointed shell.

The turret shell (*Turritella exoleta*) is the most prominent family representative in the Florida and West Indies area. It grows to three inches long, but makes its home in fairly deep water so is seldom seen except after storms.

On the western coast of Mexico, in the vicinity of Guaymas, the plentiful species *Turritella gonostoma* stages an interesting spectacle at egg-laying time. The shells wait until the tide goes out, and then, in shallow water, where they can easily be observed, they rise to the surface of the mud bottom. With mechanical precision, they line up to form a perfect circle, approximately a yard across, with the spire tips all pointed to the exact center of the circle. While in circular formation, they deposit their eggs, and then again bury themselves in the mud as the tide comes in.

Volute

Shells of the aristocratic *Volutidae* family are highly coveted by collectors. Its approximately two hundred known species are found chiefly in the deep warm waters of the Southern Hemisphere. Little is known about many of them. Some species are known to exist only because they have been found in the stomachs of deepwater fish, and nothing at all is known about their life habits. Their beauty, scarcity, and inaccessibility combine to upset the balance of supply and demand and cause very high prices.

A never-ending controversy exists over the naming and classification of volutes. Different species of shells are named, renamed, reclassified, added to, and subtracted from the Volutidae family until much confusion exists. Confusion also exists over their reproduction. A few shallow water species have been observed laying egg capsules in the cupped half of a bivalve shell, but beyond that it is mostly speculation. Many kinds of deepwater shells can be dredged with comparative ease, but not volutes. They seem to thrive in rocky areas where dredges cannot operate. Only deep-sea fish and deep-sea divers are privileged to see them.

In my collection are two specimens of the *Voluta roadnightae,* taken from the stomach of a deep-sea fish (Plate 32a). What an interesting tale they could tell. None of that species has ever been found any other way. So we can do naught but contain our curiosity while the shell and the fish share their secrets.

Another rare and valuable volute in my collection, a *Voluta* (*Cottonia*) *nodiplicata* Cox, was caught in a most interesting way (Plate 32a). A fishing boat dropped anchor for the night in deep water off the southwest coast of Australia. Next morning the shell was attached to the anchor and drawn up into the boat. The fisherman, having no interest in or knowledge of shells, took it home unaware of its great value and rarity. For two years his wife used it as a doorstop. Then, one day a friend came to visit. Being a shell collector, he suspicioned its rarity and bought it from the fisherman. Later the friend took it to an expert at the Australian Museum in Sydney for identification. Happily surprised, he learned the shell was far more rare and valuable than his fondest hope. Not only did it belong to an extremely rare species, but its size was most unusual—fourteen and four-tenths inches, instead of the usual eight. He then faced a very difficult decision. Should he keep the rare and valuable shell, or should he trade it for a whole cabinet full of more common shells. Being no longer physically able to go on shelling trips, he finally decided in favor of the cabinetful.

I, too, had a difficult decision to make. Would my project benefit most from the one very rare and valuable shell, or from the many more common ones needed to equal its value. I finally decided to sacrifice the cabinetful in favor of the one.

The 1964 edition of *Van Nostrand's Standard Catalog of Shells* lists this volute as the largest known representative of its species. There is always, of course, the possibility that the record may be broken at any time, but as far as I know no one has yet reported one larger than mine.

I have so far seen only three representatives of the species besides my own, and they are all considerably smaller. One is in a private collection in California. Another is small and of very poor quality, but it was given a place of honor in a Florida shell show. And a third, also of lesser quality and far smaller than mine, was given a special showing in the Smithsonian Institution. It was displayed all alone, in a velvet-lined case, with special lighting, and emphasis was placed upon its extreme rarity.

Junonia

The *Voluta junonia* is the pride and joy of Florida, since it is found in waters adjacent to Florida and nowhere else on earth (Plate 32c). A number of years ago it was called the peacock tail volute, and one small shell would bring $200. Later, some junonia beds were found, the market was flooded, and the price dropped drastically. The shell is just as lovely now as ever, but at today's crippled market a really good one can be bought for a far more reasonable price. It is still highly treasured, and collectors now fondly call it just junonia. A left-handed specimen was discovered several years ago. Valued at $1,000, its fame spread far and increased the value of all junonias for quite some time.

The junonia grows in deep water, and must be brought up by divers. Occasionally, after a storm thoroughly churns up the waters, a live one is washed, undamaged, on the beach. In 1951 it was my good fortune to be on exactly the right Florida beach at exactly the right time to make such a find. The rarity of the event was impressed upon my mind when word rapidly traveled over the area that an Indiana woman had found a good junonia. Ten years later, in 1961, it was again my good fortune to find a second good one washed ashore. I showed it to a collector whose home fronted on the beach and also told her of my find ten years earlier. She said: "And here I live right on the beach, search it every day, and I have never even found a dead one."

The junonias found along Florida's central west coast, those found farther north nearer the Mississippi River mouth, and those found farther south in the Campeche Bay area, all differ in background color. The trio makes an interesting study in a collection.

New Zealand Volute

Another volute in my collection of which I am justly proud is the much sought after golden volute of New Zealand, the *Voluta* (*Iredalina*) *aurantia* Powell. Through the help of New Zealand friends, I was able to add one to my project. It inhabits very deep water, and very much resembles the famous golden cowrie in color, although it lacks the extremely high gloss of the cowrie.

Northwest Australia Volute

The striking black-trimmed *Voluta bednalli* of northwestern Australia is another highly coveted species (Plate 32 a and d). Although quite small, the color pattern is so unusual as to greatly enhance any volute exhibit. So great is its demand that, even on location in the remote area where shells are usually cheap, it sells at $100. The one in my collection (see photo) is of extra special quality and very beautiful.

Music Volute

The *Voluta musica* Linné of the West Indies is perhaps the most unusual in appearance of all volutes. A trim pattern circling the shell exactly resembles the bars and notes of printed music. It has been observed gluing small, oval, flattened egg capsules in the cup of bivalve shells.

Voluta Magnifica

One of the largest and most distinguished species is the *Voluta magnifica* Geb. of eastern Australia (Plate 32e). It lives buried in sand and mud just beyond low-tide line and can often be collected by wading at extreme low tide. Although it grows to twelve inches and is no less beautiful than the deepwater species, it is far less expensive since it is more easily collected and, therefore, more plentiful.

Imperial Volute

The *Voluta imperialis* Lft. which inhabits the Philippine Islands, is another volute much in demand by collectors (Plate 32f). It is also called the Chinese emperor's crown and grows to eight inches in length. The distinctive crown of spines, salmon-colored lip, and background color make it a lovely and colorful specimen for a cabinet.

The genera *Melo* (see page 142) and *Cymbium* also belong to the family Volutidae.

Wentletrap, or Staircase Shells

The approximately two hundred different species of wentletrap, or staircase shells, are widely distributed from arctic to torrid seas of both hemispheres, and from shallow water to the greatest ocean depths. They are also often taken from the stomachs of fish. The shells of these small creeping mollusks, which exude a purple fluid to confuse an enemy, are decorated with a spiral flange remindful of a winding stairway. The term *wentletrap* is a corruption of the German word *wendletreppe*, meaning winding stairway.

The two-and-one-half-inch precious wentletrap (*Epitonium scalare* Linné) is the species for which the family is most renowned (Plate 33). It was especially favored by the royalty of Sweden, Russia, and Austria. In ancient times an emperor paid the equivalent of $20,000 for a single specimen.

The shell was first known from specimens collected in Indonesia. Later it was found in Ceylon, Japan, and China, and still later, in Australian waters. However, the shell is most associated with China.

About 1700, this shell was still so

scarce and in such demand that a single specimen sold for the equivalent of $200. According to an unverified story, the clever Chinese then learned to mold a counterfeit shell from rice paste, and the world was flooded with perfect imitations. The real precious wentletrap soon became almost valueless, and their price has never significantly increased; for enough of the shells have now been found in new and different areas so that really good ones can be bought for modest sums.

Whelk, or Conch

There is as much confusion and variation in the application of the common term *whelk*, as with the use of the term *conch*. They are each used for several different families, and sometimes both applied to exactly the same shell (Plate 34a).

Left-Handed Whelk

The so-called whelk that is perhaps most widely known is Florida's common left-handed, or lightning, whelk—*Busycon contrarium* Conrad (for many years known as *Fulgur perversum* Linné). It was caught up in one of the many scientific shell-naming controversies. Exactly the opposite to most other shells, its large whorl and opening is to the left of the central axis, or pillar, known as the columella.

I felt there must be some scientific explanation for the unusual left-handed formation and, as so often happens, I received conflicting answers to my inquiry. The explanation that somehow seemed to me to be most logical I offer below, although I know it is held in dispute.

On rare occasions a *right*-handed left-handed whelk is found. It is of considerable value because of its rarity, whereas the normal left-handed whelk is common, plentiful, and of no special value. A matched pair is even more precious. I have the most perfectly matched pair that it has so far been my privilege to see—found alive on the same day within three feet of each other on the same beach near Sarasota, Florida (Plate 34b). They are so completely identical it would seem they must certainly have been blood brothers. I have been advised to place a value of at least $100 on the pair because of its rare perfection. Had they both been left-handed, the two would be worth no more than $1.50.

At low tide, the left-handed whelk is often found burrowing with its broad foot just beneath the sand surface in search of bivalve food. It is believed that it is guided by its keen sense of taste to the streams of water being expelled by the clams it seeks. The siphon tubes through which water circulates to the gill chambers extend above its protruding head. Waving tentacles, each bearing a little black eye, come out of each side. It is equipped with a boring, sucking proboscis capable of dealing death to any bivalve unfortunate enough to cross its path. The small- to medium-size whelk kills its prey by grasping the clam in its large foot and holding the valves together to smother it. The large whelk can open a clam by gripping it firmly in its foot, then, by a sharp contraction of the foot muscle, hammer the clam open with its own heavy shell. This habit accounts for many of the broken edges on otherwise good live-taken whelks.

As the whelk goes about the business of hunting its food mollusks, it is, in turn, being hunted by its own enemies. Lying in wait for it, concealed from its view in rock caves and crevices, are the ever-hungry stone crabs with massive purplish bodies. With their strong, brightly colored claws, they can break away the whelk's shell, piece by piece, as they feast upon the juicy meat within. That same juicy meat is much in demand by the sea gulls flying overhead, those clever birds that somehow know by instinct how to cope with a seemingly impossible problem. Having no equipment of its own with which to crack the shell, the gull grasps the mollusk in its claws, sails high into the air, and drops it to the ground. The gull follows it as it falls, and enjoys its feast among the broken fragments of shell.

The whelk's eggs are in small, round, flat capsules, a little larger than a quarter. They are attached close together in an egg ribbon, sometimes a yard or more long. The female whelk attaches the first end of the string to an old shell, or something else that can serve as an anchor. The first few capsules are undersized and widely spaced. Then follows the perfect long chain, or ribbon. I have never personally counted the eggs in my specimen, but the statistics from a museum case where I saw a similar ribbon displayed were 165 capsules, 80 to 90 eggs to each capsule, or about 13,000 eggs in the one egg-case ribbon. The embryo is nourished in the capsule. At hatching time a small section of each capsule drops from one side and the tiny infants, each exact replicas of the parent shell, fall out to begin their independent life on the sea bottom. Of those thirteen thousand tiny babies, it is estimated that only one in ten thousand lives to perpetuate the species. The huge surplus is nature's way of assuring the continuity of the species, as well as ample food supply for her ocean creatures.

Large whelks were used by the Indians for drinking vessels, and the sharp lip edge was made into cutting tools. Now, in modern times, they can still serve many functions—borders for paths and flower beds, flowerpots and lamps, to name but a few. They are much in demand as trade shells in the islands of the South Pacific. Many of the natives there believe their family will be assured of fertility if they own a left-handed whelk. I have a shell-collecting friend who has had amazing success in the South Pacific trading with Florida's left-handed whelk, since he learned of this superstition.

European Whelk

The red whelk (*Neptunea antiqua* Linné) of northern Europe is perhaps the most widely known, and used, whelk of that area. It is reddish in color, from three to six inches long, quite plentiful, and is used extensively for both codfish bait and human food in the Great Britain area. It is sold in different markets under the names of barnagh, almond, and buckie, and is considered by many to be even more delicious than lobster.

New England Whelks

Two large whelks of special interest are found along the coast of New England. One, the knobbed whelk, has long been a scientific bone of contention. For many years it was listed under the Latin name *Fulgur carica* Gmelin, but now, after much dispute, its name has been changed to *Busycon carica* Gmelin. The nine-inch shell has a beautiful reddish interior, and the animal inside is very large—and very heavy. I can truly vouch for the latter. During a shell-hunting expedition on Cape Cod in 1951—and while at least three miles down the beach from my car—I waded far out at low tide in a mud flat area. At first I thought I was wading among stones about the size of small grapefruit. Then I saw one move and suddenly realized I was in the middle of a large colony of partially buried knobbed whelks. In no time flat, and with the greatest of ease, I had my large kettle full to overflowing without even making a dent in the supply. But carrying the load three miles back to the car—well! That took a bit of doing.

The second large North Atlantic whelk found along the New England coast, the channeled whelk, was also under scientific dispute for many years. Formerly listed as *Fulgur canaliculatum* Say, it now answers to the name of *Busycon canaliculatum* Linn. This lovely, also nine-inch, shell has an especially rich, velvety, glossy interior. The early Indians cut sections from this shell into beads to make their wampum belts. Three beads were worth a penny in early colonial times, and a five-foot string was worth five shillings.

Polar Area Whelk

The waved whelk (*Buccinum undatum* Linné) has adapted itself to a wide range of temperatures. Some are found in temperate areas, but it does particularly well in the polar regions. This shell, which grows to anywhere from three to six inches, is an important source of human food in the far north. The meat is boiled until tender, fried in fat, and then eaten with oil and vinegar or put into a vegetable soup.

The waved whelk plays an important role in the lives of both lobster fishermen and cod fishermen of the far north—hindering one and helping the other. These whelks travel in droves and are aggressive, carnivorous, heavy feeders, and their organs of smell are developed to an unusual degree. The combination is a fatal one for lobster fishermen. Smelling the bait in the lobster traps lowered to the bottom, the whelks approach en masse and soon devour all the bait. But, while lobster fishermen rue the day whelks were created, cod fishermen are grateful to them. Codfish are especially fond of whelks; often forty or fifty of the shells are found in the stomach of a single codfish. So great quantities are used as bait by cod fishermen.

The far north waved whelk, contrary to other whelk species, lays its egg capsules in a large mass resembling a coarse sponge, and not in the more typical long cord, or ribbon. Many years ago soldiers used these "sponges" as a soap substitute, calling them sea-wash balls. Each egg capsule looks like a large split pea, and there are five hundred or more in one mass. As they hatch, the stronger ones eat the weaker ones—just one more of nature's unique ways of providing food for her infants.

12. Land and Tree Snails of Africa, France, and Florida

Our discussion of mollusks is not complete without also including land and tree snails. They are found in the cabinets of collectors, just as treasured as the mollusks of the sea. And, in life, they more closely affect the lives of people than do their water-loving cousins. So with these next few pages, we will give a little history of the land cousins, and thus complete the picture of how the lowly mollusk manages to survive in both environments.

These odd little creatures are air breathers and, as everyone knows who has seen a snail crawling about in his garden, they extend a long slender foot from the shell to crawl, gracefully gliding along as tiny muscles contract in a rhythmic succession of waves to propel them forward. The eyes are the little bulges at the tip of the feelers (called eyestalks), extending from the head at the end of that crawling foot. Thus, they differ from most water snails, where the eyes are at the base of the feelers, but the radula, or rasplike tongue, for tearing food, is common to both. The two shorter antennae by the eyestalks are extremely sensitive organs of touch.

Most eat vegetation, foraging widely by night, then resting by day in holes among the grass roots. They conceal themselves well by cementing a roof above themselves with leaves and rubbish. Then they retreat into the safety of their shell and, with a gland in the foot, secrete a porous stopper.

They have a great capacity for sleep. Besides several months of winter hibernation, the snail crawls into its shell during weather extremes. It takes short naps or spends days curled inside its tiny home as a protection against hot sun or heavy rain. Desert snails can, miraculously, doze off for as long as three or four years. There are records of land snails surviving under extremely adverse conditions. The most well-known instance is that of a *Helix desertorum* land snail from Egypt. It was found to be still alive after having been attached for four years to an exhibit table in a British museum. After removal it revived and lived through two summers.

When weather conditions are right, the snail eats ravenously. Being extremely nearsighted, it must depend upon its strong sense of smell to guide it to its food. Its tiny mouth is equipped with some 25,000 extremely small teeth, which are able to grind through almost anything with which they come into contact. Many collectors have learned to their sorrow that if this snail is put in a cardboard box, it will very quickly chew its way to freedom. When the teeth are worn down, the snail replaces them as quickly and easily as it can repair a broken shell.

Each spring and fall, instinct tells the snail that it is time to perpetuate the species. It loses interest in food, its antennae extend full length and probe about in all directions. Suddenly, its decision made, it takes off at top speed on a miraculously unerring course to the lady of its choice. The courtship lasts several hours and the pregnancy lasts from twelve to fifteen days. The pregnant snail then lays up to twenty-six eggs, always choosing a spot where the food supply is right for newly hatched infants.

The eggs are round, and in some species almost transparent, whereas other species bear eggs coated with a limy substance that leaves them opaque. They are up to three-quarters of an inch in size. Some are deposited on the ground surface, while some are buried beneath leaf mold. Some are attached in clusters to the underneath side of leaves, and some snails seal the leaf edge over the eggs to form a pouch. For its first meal, the newly hatched snail feasts upon the eggshell from which it just emerged.

Searching for the timid little creatures is not exactly easy since you must first discover their hiding places. The collector must search on leaves, branches, tree trunks, bushes, in grass and weeds, around and under stumps, stones, logs, under leaves, along creek banks, in holes in the ground, and so on. In general they prefer moist areas.

Let us look briefly now at a few species of particular interest.

Large Land Snail of Africa

Africa, of course, has many different species of land snails, but the most well-known species is the *Achatina fulica*, which reaches a length of eight inches or more (Plate 23b). Only Liberia's *Achatina achatina* is larger.

By various means this large *fulica* snail has now migrated to several different areas of the world. Because of its huge vegetarian appetite, it soon becomes a serious pest in agricultural areas. It was my privilege to see a large truck farm that had been invaded, and I was truly amazed at the near total destruction it can cause.

The most recent alarming invasion of this snail has been to the United States. In 1966 a young boy became attracted to them while visiting in Hawaii, and brought three back to Miami, Florida, as a present for his grandmother. The warm climate and lush vegetation was ideal for their development.

They lay small round white eggs, and each snail is capable of producing six hundred offspring each year. By the fall of 1969 they had infested a thirteen-square-block area in the Miami area. Alarmed, agriculture department officials investigated and estimated that there were at least twenty thousand in this one small area. The snails were consuming everything in sight and were even climbing over houses, eating off the paint to satisfy their craving for calcium.

Fully aware that the entire South Florida agriculture area is now in grave danger, the authorities are concerning themselves with a problem for which there is no sure solution.

Experiments at control have been carried out in other countries and the best method seems to be the use of parasitic snails that live upon and kill the young *Achatina.*

Florida Tree Snail

The beautiful *Liguus* tree snail for which Florida is noted is found only at the southern tip of the state (Plate 23c). It is believed to have been blown from Cuba by hurricanes thousands of years ago and is found nowhere else in the United States. Civilization has encroached upon its domain until it is now near extinction, except within the area of the Everglades National Park where it is under government protection.

A desperate attempt is now being made to preserve all still existing species that are in grave danger of being crowded to extinction by modern development. Dedicated conservationists are collecting the remaining (near extinct) species from outlying areas and transplanting them to a like environment within the confines of the protective national park. This operation is proving quite successful now that they have learned to choose the right area. Many first attempts were a failure as many people tried to move them to areas unsuitable for survival.

These snails live high on the branches and trunks of the gumbo-limbo and Jamaica dogwood trees. The host trees are in no way damaged as the snail's entire food consumption is the lichens, moss, and tiny plant life growing on the tree bark. In fact, much experimenting is now being done in an attempt to use snails as a control for the harmful fungus growing in orange groves.

The *Liguus* is being very carefully studied, with continuous new findings. It is believed that its habit of living its entire life in one tree contributes much to the ease with which it can be hunted to extinction. Many times it is knocked to the earth by heavy winds, but it is believed that it never voluntarily leaves the tree to which it is adapted except to lay eggs. Each August or September it digs in the leaf mold at the base of its own special tree to lay eggs. Over a period of several days these snails lay as many as fifty small brown eggs on the ground beneath the mold. The eggs lie there for approximately eight months awaiting the spring rains before they hatch. How the tiny babies are led by instinct up the trunk and into the tree is but one more of nature's marvels.

Snails are often found in groups in the tree. They spread a slimy film on the limb ahead as they crawl, and when others happen across this ready-made walk they follow it. When disturbed by a threat of danger, they have been observed to exude a surprisingly large quantity of the sticky fluid. With careful observation, it has been determined that their rat and bird enemies find this fluid extremely distasteful and drop the shell that they had been intending to eat.

It is estimated that the life of a *Liguus* tree snail may occasionally extend up to nine years, although three to five years is considered more normal.

According to Treat Davidson, in his article in the March 1965 *National Geographic* magazine, each fall, after the egg-laying is completed, the snails return to a tree limb and anchor themselves with a mucus substance. It hardens and holds them firmly in place for approximately eight months of winter hibernation. The hardened mucus not only anchors the shell, but it also thoroughly seals it to conserve all moisture inside. With the coming of the spring rains, the mucus is dissolved to release the shell, and it resumes its eating and growing.

This is another of the areas where conflicting information makes the writing of this book most difficult. After fully believing Mr. Davidson's version since his 1965 report, I am now told by an equally respected authority at the Miami Marine Laboratory that the *Liguus* hibernates in the leaves at the base of the tree.

With regard to activities hidden beneath water and sand, it is easy to see how differing beliefs among researchers might possibly *never* be resolved to the satisfaction of all. But, in the matter of the *Liguus* snail, whose activities are exposed to direct observation, there would seem reason to hope that, in time, all will arrive at an unconfusing single belief.

It was my rare good fortune to be included in one of the controlled *Liguus* collecting trips now permitted in the Everglades area. To reach the interior hammocks (or islands) where the elevation is high enough to permit tree growth, one must cross large swampy areas of saw grass. Of course, no roads are available, so air-driven boats or swamp buggies, equipped with many huge balloon tires, must be utilized. I will only state here that anyone contemplating such a trip should first carefully examine the strength of his motivation. Physical exertion, hot sun boiling down, biting insects by the million, snakebite possibilities, thirst, and so on, all combine to test even the strongest desire.

French Edible Land Snail

Among the many edible species of land shells, the *Helix pomatia* is most noted for its excellent flavor.

Edible snails of the *Helix* genus thrive and are eaten in many areas of the world. East Germany has a big snail farming business, but France has grown most famous for its snail consumption and commercial snail farming. As many as one hundred thousand are consumed daily in Paris alone (twenty snails to the serving).

France also imports large quantities of live edible land snails from other countries to satisfy its need. Through these live imports, many different species have now established themselves there. In one instance a twelve-ton shipment from Turkey was pronounced mostly dead upon arrival. Eight and one-half tons were dumped in the forest. A few days later they arose from the dead to continue their life, and reproduction, in France. The yearly consumption in France is estimated to be some 250 million. And with snail eating becoming increasingly popular in the United States, France annually exports some 24.8 million of the delectable creatures to us.

Small border wars are often caused by controversies over snails. One such incident took place in April 1965 between France and Switzerland. The Swiss place more stress on the beauty of their countryside for both themselves and tourists than the French do. The French take great delight in going across the border into Switzerland for picnics, where the main event is collecting snails and roasting them over open fires. They often create great damage, and Switzerland passed laws to end this French practice. Then, in retaliation, in an atmosphere of ill will, the French passed laws forbidding the Swiss from continuing to cross the border to collect wild flowers, wild strawberries, and mushrooms in France.

So popular did snail eating become that demand exceeded supply, necessitating snail laws. In the area of Neuchâtel, Switzerland, one must now buy a license to hunt snails. With the license, the hunter is given a small metal ring 1.3 inches in diameter, and he does not dare keep a snail that is small enough to slip through this ring.

Medical research is now also making substantial inroads on the snail supply. The University of Paris buys several hundred pounds each spring for experimental use. The gastric juices are considered invaluable in the study of sex and adrenal hormones in the human bloodstream. It is hoped that with further experimentation the juices will make it possible to detect the early presence of certain types of cancer.

We will now leave the shell-bearing mollusks of both sea and land and make the acquaintance of the perhaps even more fascinating shell-less mollusks of the sea.

13. The Many-Tentacled Cephalopoda

The familiar squid and octopus, and their somewhat less familiar cousins nautilus, cuttlefish, and spirula, are so vastly different in appearance from clam and snail mollusks that it is difficult to believe they are mollusks at all. However, they all have soft unjointed bodies that are characteristic of the mollusk classification, so indeed they are mollusks. Their scientific name is the jawbreaker Cephalopoda. *Cephalo* means "head" and *poda* means "foot," so they are referred to as the head-footed group.

The head of each cephalopod is circled with long arms, or tentacles, equipped with suction disks, and a mouth with a horny parrotlike beak lies in the center of the animal. Inside the beak is a flat ribbonlike tongue equipped with many fine sharp teeth that are most effective in tearing apart a meat meal. Each expels water from a funnel-shaped mantle cavity to propel the body backward through the water. They all breathe through gills. Most, except the nautilus, are equipped with ink sacs. With the exception of the nautilus, none has an external shell, although the squid, cuttlefish, and spirula have internal shells. Their skin contains pigment spots by which they can change the color of their bodies at will. They have a brain, and all have well-developed eyes, organs of smell and hearing, and a complex reproduction system.

Through the study of fossils, geologists have determined that many, now extinct, species of the Cephalopoda existed in prehistoric times, including many of giant size. Fossils of the ancient ammonite, an animal similar to our present-day, much smaller, chambered nautilus, have been found with a diameter of four feet.

Chambered Nautilus

Of the five Cephalopoda cousins, the chambered nautilus (*Nautilus pompilius*) fits most closely into the seashell image (Plate 34c). Because of the beautiful material in its external shell, it is called the pearly nautilus. This mollusk is found only in the warm waters of the South Pacific and Indian oceans, and is one of but a very few living representatives of its once numerous and widely distributed family.

The growth of the shell of the chambered nautilus, and the living arrangement inside, is perhaps the most unique in the shell world. When first hatched, the tiny baby fills its tiny shell just as any other univalve mollusk. But when the soft body and the shell covering both grow a size larger, it seals off a small empty chamber and lives only in the forward part of the shell in front of the pearly partition. From that empty chamber, through the walled partition, it builds a delicate hoselike tube called the siphuncle, connecting it with the front-occupied chamber. Then when it grows another size larger it seals off another empty chamber, extends the siphuncle tube, and moves another space forward in the enlarging shell. This process is repeated over and over as growth continues. Often the adult mollusk, still occupying only the last and largest space, has as many as twenty-five or thirty sealed-off chambers in back of it, and they are all connected to it by the siphuncle tube. There is still scientific disagreement as to exactly how that tube is used. Some authorities believe it admits air into the sealed-off chambers so this mollusk can carry its shell with ease as it walks on the ocean floor. Another theory is that through the tube a gas is secreted and forced into the chambers. It is believed that, by controlling the gas pressure inside the chambers, the shell can gain, or lose, buoyancy to rise or sink for food or safety.

There is scientific disagreement as to the habits of the nautilus. It seems most generally believed that it forages energetically on the bottom of the sea, possibly at a depth of two hundred feet or more. It is known to have an up-and-down vertical migration, rising each night for a time to within two hundred feet of the surface. In the area of New Caledonia it has been known to venture into shallow water, where it has been taken by divers. And at times, attracted by bait, it has crawled into fishermen's traps. Though it is one of the oldest animals still living on the earth, its actual living habits, reproduction, and so on are still a mystery.

The eyes of the nautilus have no lenses. It is one of the few examples in nature where visual organs operate as does the pinhole camera. The soft-bodied animal has many arms (approximately ninety in the female and sixty in the male). The mouth is concealed at the swollen base of these tentacles. When the tentacles grab a crab or other live animal, the head protrudes, two horny parts of a beak tear the body and cut it like a pair of scissors, and the rough tongue carries the chunks of meat into the digestive cavity. While swimming about in search of food, the soft animal protrudes far from its shell. But, in times of danger, it withdraws completely within it.

The nautilus is often strung on the roofs of Polynesian homes in the belief that it possesses a mysterious power to chase away evil spirits. The meat is widely used in some areas for food, and the beautiful pearly shell adapts to many uses. Dippers, vases, spoons, buttons, and many ornaments made from the nautilus are widely sold in curio stores. In Paris quantities are used for the finest of cameo cutting. In India it is ornamented and used for drinking cups. In America it is highly valued as a cabinet specimen, both in the natural state, and polished to reveal the pearly interior.

This quite fantastic shell has been immortalized in the minds of men by this famous poem of Oliver Wendell Holmes —a poem in which he felt inspired to describe one of God's masterpieces:

The Chambered Nautilus

This is the ship of pearl, which, poets feign,
Sails the unshadowed main,—
The venturous bark that flings
On the sweet summer wind its purpled wings
In gulfs enchanted, where the Siren sings,
And coral reefs lie bare,
Where the cold sea-maids rise to sun their streaming hair.

Its webs of living gauze no more unfurl;
Wrecked is the ship of pearl!
And every chambered cell,
Where its dim dreaming life was wont to dwell,
As the frail tenant shaped his growing shell,
Before thee lies revealed,—
Its irised ceiling rent, its sunless crypt unsealed!

Year after year beheld the silent toil
That spread his lustrous coil;
Still, as the spiral grew,
He left the past year's dwelling for the new,
Stole with soft step its shining archway through,
Built up its idle door,
Stretched in his last-found home, and knew the old no more.

Thanks for the heavenly message brought by thee,
Child of the wandering sea,
Cast from her lap, forlorn!
From thy dead lips a clearer note is born
Than ever Triton blew from wreathèd horn!
While on mine ear it rings,
Through the deep caves of thought I hear a voice that sings—

Build thee more stately mansions, O my soul,
As the swift seasons roll!
Leave thy low-vaulted past!
Let each new temple, nobler than the last,
Shut thee from heaven with a dome more vast,
Till thou at length are free,
Leaving thine outgrown shell by life's unresting sea!

Paper Nautilus

Much confusion exists between the chambered nautilus and the fragile paper nautilus (*Argonauta argo*) (Plate 34d). Because the shells of both cephalopods are the same shape, many people assume that the thin fragile shell of the paper nautilus is just a young, not fully formed, chambered nautilus. However, such is not the case. They are two entirely different animals with wholly different life habits.

Knowledge of the existence of the paper nautilus dates back to antiquity, but an understanding of it has prevailed only through the last century. According to belief (stated by Aristotle) fleets of them were seen floating in the Mediterranean area in ancient times. Early Greek and Roman writers called them Argonauta, alluding to the Argonauts who sailed away to find the Golden Fleece. They were given the name *nautilus,* which means "little sailor." To distinguish them from the pearly chambered nautilus (with its similarly shaped but heavier shell), they were called the paper nautilus, an apt name since their shell is very nearly paper thin and extremely fragile. As late as the middle of the nineteenth century, little more was known about them than the writings of Aristotle four centuries before Christ.

The paper nautilus animal did not permanently wear its shell in the usual manner of mollusks, but was able to leave it entirely at will. So it was assumed that the animal had no shell-secreting glands, and that it was actually some kind of hermit crab occupying the vacated shell of another mollusk. The question of how the shell was formed remained a mystery for centuries. It was not known whether the argonaut could survive without the shell, and no one had ever seen a male argonaut. By the middle of the nineteenth century, it was still believed the animal was incapable of secreting a shell. Then still another mystery arose. A long squirming object resembling the arm of a tiny octopus was found in the animal's mantle. This squirming object was found to be full of spermatozoa—the male element for fertilization.

Intensive study followed this discovery. Eggs were hatched under control for observation. Little by little, bits of information fell into place. The best that I can learn with the researching I have done indicates the mystery was finally solved in 1853.

Usually the shell-secreting cells of a mollusk are found in the mantle. The mantle of the argonaut had none. Then it was discovered there were shell-making cells in two large webs that spread out at the ends of its two rear (dorsal) arms.

When the young female is but one inch long she begins to form her paper-thin shell. Her body rests in the mouth of the shell with most of it extending outside. The two inflated webbed arms continuously and tightly clasp the two sides of the shell, making constant deposits of shelly substance to increase its size. Often the fragile shell is seriously broken and the webs do an amazing repair job. There is no connecting tissue of any kind between the shell and the animal. The arms simply hold it in place at all times, and they hold it in such a position that the eggs are laid in the back coil of the shell where the mother's body in the shell opening gives them protection.

The male of the argonaut grows to the tiny size of but one inch. At mating time one of his eight arms is modified into a sex organ capable of producing the male element for fertilization. It was determined that the slender object long observed in the female's mantle was indeed the detached fertilizing arm of the male of the species.

After the eggs are laid and fertilized, a

gland in the female's body then secretes a sticky substance over them which hardens and draws them together in a cluster like a bunch of grapes in the shell back of her, where they remain until the young hatch and swim off as free individuals.

When the nautilus reaches the end of its life, the paper-thin shell, so highly treasured by collectors all over the world, is then entirely at the mercy of the ocean waves. It may be thousands of miles from the nearest beach. Its chance of reaching a beach is identical with that of a floating cork. And its chance of making a safe landing is identical with that of an empty eggshell landing unbroken, thus making understandable the great thrill of finding one in perfect condition. It also explains why a single good specimen sometimes sells for an amazingly high price.

In 1963 it was my extremely good fortune to get a good supply for my collection. Just before I reached North Island in New Zealand, one of those occasional, once-in-a-lifetime phenomena occurred. Storm winds blew from exactly the right direction to strand a huge school of argonauts on a little island just off the coast. A friend of mine happened to be on the island shelling when they came in. He had until the next high-tide hour to rescue as many as possible. Many of the fragile valuable shells were broken. With a few frantic hours of work racing the tide—hours that he can certainly never forget—he managed to rescue approximately fifteen hundred unbroken specimens. He had them all nicely cleaned when I arrived, and it was truly a beautiful sight to behold. I have a beautiful growth series in my collection from those fifteen hundred.

Spirula Spirula

Unlike its nautilus cousins, the *Spirula spirula* Linné carries its beautiful shell inside its body (Plate 34e). Its beauty cannot be enjoyed until the animal dies. The shell is a tiny pearly tube, little more than an inch across, coiled in the shape of a ram's horn. In the tube are many cross partitions made of a material that glistens like jewels. It is the only member of its family and is found in warm and temperate seas circling the globe. Like all its cephalopod relatives, it has a soft body with long tentacles circling the mouth at one end. The rear bulging end is most unusual. It emits a yellow green glow, and buried within it is the beautiful little coiled jewel of a shell. Since the small coiled tube is hollow, its buoyancy upends the animal so that it must spend its entire life hanging head down in the water.

These small animals live in great numbers in the open sea beneath the zone of light penetration. When they die and decompose, the small coiled shells ride the waves and ofttimes land unbroken on a beach far from where they had lived. Finding a perfect one always brings joy to the heart of any collector. They are reputed to be plentiful on most tropical beaches, but, personally, I did not find it so. Usually, when I found one, that was it. Just one. In my many years of searching, I found but one beach area where they prevailed in great numbers. On the west coast of North Island in New Zealand, I collected several hundred perfect ones on a strip of beach no more than a mile long. Apparently, ocean currents and prevailing winds connected that beach area perfectly with some thickly populated section of the ocean.

While I was in New Zealand in 1963, I was told that down through the years, countless millions had washed ashore annually on their beaches. I was also told that, for many years, it remained a mystery where they came from since no living creatures could be found. Then, finally, the Dana expedition, under the late Professor J. Schmidt, succeeded in trawling a number of them alive. Professor Schmidt found them in the waters of the North Atlantic at a depth of from 110 to 1,100 fathoms (6,600 feet).

Since that time, Dr. Gilbert L. Voss and his associates, of the Institute of Marine Science in Florida, have taken many at a depth of approximately sixteen hundred feet.

Cuttlefish

The cuttlefish (*Sepia officinalis*) (Figure 11) is a small Old World inhabitant of the Cephalopoda mollusk group. It is closely related to the octopus and squid, although it grows to but a few feet in length.

Comparatively few people have ever seen one—which is their loss—since the animal is a strikingly showy object, beautifully banded, mottled with black, brilliantly iridescent in the sun, and quick at changing color by the manipulation of pigment cells in the skin. The body, molded around the broad internal shell across its back (the cuttlebone used in birdcages all over the world to feed canaries and other pet birds), is broad with flat narrow fins edging it all the way around. The head is at one end, with two long flexible tentacles on either side of two large black eyes. And it wears a crown of eight short arms and two long ones, each studded with rows of suckers. Naturalists deem its arm motion to be the most rapid known in the animal kingdom—even more rapid than the tongue of the toad or lizard. It is truly a beautiful and unusual product of nature.

For its own protection, it is equipped with an ink sac with which to cloud the water in front of a pursuing enemy. Although the sac of sepia, or ink, is rarely larger than a man's thumb, the force exerted upon it throws a jet stream of ink for six feet through the water, coloring a large area black. The ink from this mollusk is the original India ink, and it is still valued highly by modern artists.

Squid

The squid, like the cuttlefish and octopus, is a highly developed mollusk of the Cephalopoda class. It has the characteristic soft-tube body with the head surrounded by tentacles—ten—two of which are much longer than the others. Unlike the cuttlefish, the horny penlike shell beneath the skin of its back is narrow rather than broad. When pursued, it, too, like the cuttlefish and octopus, ejects an inky fluid that assumes odd forms in the water to conceal its movements. The many species of squid range in size from nearly sixty feet long to a tiny midget species that will not cover a dime.

Figure 11. A live cuttlefish, showing the tentacles. *Rex Gary Schmidt, courtesy U. S. Fish and Wildlife Service*

There are many species of squid, and their reproduction process results in a most unusual commodity. The female's egg case is called a sea mop and is used by natives as a substitute for soap. It is composed of long banana-shaped sheaths, each containing hundreds of eggs, approximately forty thousand in one mop.

Hatched from the egg cases, the tiny babies look like soft lumps of clear semifluid jelly. By far the majority are devoured in infancy by fish, and a very small percentage ever reaches maturity. They are great consumers of the smaller fish species; and they, in turn, are one of the major foods for large animals, such as the dolphin, porpoise, and toothed whale.

Nature really went all out mechanically when she created the squid. The principle behind the piston-vacuum action of its suction cups is used by man to solve many of his problems. And the principle of jet propulsion was perfected in the squid, and later copied by our airplane designers.

The squid is equipped with side fins for slow motion, but when it is in a hurry, it goes by jet. It draws water into a large body cavity, or mantle chamber, and then forcibly expels it through a tapered funnel, or nozzle. The nozzle is equipped with a valve to prevent backfire, and it can be turned forward, backward, or even sideways to guide direction. Some species can even jet themselves above the water surface and sail through the air for short distances. The squid is one of the swiftest swimmers of the sea, with its great speed originating from an unusual set of nerves in the mantle. These nerves are composed of giant fibers one hundred times the size of those in man, and their strong impulse to the muscles of the nozzle causes a quick unison contraction that expels the water in a rush to quickly jet the animal away to safety. By a careful study of these huge nerve fibers of the squid, scientists have greatly enlarged their knowledge of how the human nervous system works. This massive inner signaling system of the squid is the largest known to exist in the animal kingdom and is consequently the most easily studied.

Much has also been learned about the human eye from a study of the squid's eye. It is amazingly like that of a human, although in some respects it is even more complex.

The squid's eye has an eyelid just as a human's, and it is composed of a cornea, pupil, lens, and retina. The greater complexity lies in the construction. It is constructed in a way to eliminate the blind spot of the human eye. The squid's two eyes see independently of each other. And the lens is more complicated. In the human eye the lens is focused by the eye muscle without changing its position. In the squid it moves in and out on the order of a camera lens. It even tilts at an angle so the animal can scan a wide area without moving an eyeball.

In some species the two eyes vary greatly in size. It is believed the smaller one is used when feeding near the surface, and the other, much larger one, is used to let in more light while feeding in dark deep areas.

The squid not only has the ability to change color for camouflage protection, but it also has the amazing ability to light up, something on the order of a firefly. It is universally believed that the color change is used as protection from enemies, to court a mate, and to hunt prey. But there is no scientific agreement as to the use of the lights. Some scientists believe they are used as lanterns to light the way. Others think they are to frighten away predators, or perhaps to attract food animals. Still others think the major purpose of the lights is to keep the squid together in schools in the dark ocean depths.

Some squid have hundreds of these lights (called photophores), which they wear like jewels as they shine through color filters. In some species they are mere pinpoint size, and in others they are as large as fifty-cent pieces. Using its complicated nerve impulse system, the squid can turn its lights on at will by producing some chemical change in the light organs, which is not yet fully understood.

One small squid (*Heteroteuthis dispar*) has a large luminous organ located adjacent to the ink sac. When disturbed, both organs function simultaneously and instead of ejecting black ink, the squid leaves a trail of blinding liquid fire to confuse its pursuer as it darts off to safety.

The deep-sea diadem squid (*Lycoteuthis diadema*) has an especially complicated lighting system that produces brilliant lights of three colors—two shades of blue and one of red.

Another species of special interest is the Japanese firefly squid (*Watasenia scintillans*), which is caught in immense numbers off the west coast of Japan in the spring and early summer. It emits brilliant flashes in the dark from a multitude of light organs located at various points over the body.

Besides the light organs (photophores), the squid's body is also covered with color organs or elastic pigment sacs (called chromatophores), which it can operate at will. Although each species carries but three basic colors, by opening and closing differing numbers of each it can create a wide variety of hues. Each pigment sac is surrounded by a circular muscle. And with the vastly complicated connecting nerve impulse system, the squid can quickly relax or tighten those muscles to either reveal, or conceal, the pigments within the sac controlled by each muscle. Thus, with great skill, it produces any color shade needed at the moment. The squid can make the changes so rapidly as to make waves of color seem to flow like tiny waves flowing over a windblown wheat field.

Besides the color change for protection, the squid also has an ink sac safety device. Little is known about its chemical composition, but it somewhat resembles the pigment in the hair and skin of man. Its color varies from black to light brown. When endangered, the squid shoots out a mass of ink, which assumes a size and shape in the water similar to the animal itself. It then coagulates in the water to maintain this deceiving shape. At the same instant the squid closes all its color and light organs to become a neutral color. Then it darts to one side while its pursuer grabs nothing but a mouthful of ink.

Unfortunately, squid cannot be kept long in captivity, so few people are privileged to see these odd and fascinating creatures. When placed in a tank they dart about with such speed as to often fatally injure themselves. They also eject their ink in such quantities as to suffocate themselves many times. Many just seem to die of shock and exhaustion.

Living at great depths, rarely does a really giant squid come near enough the surface to be seen by man. According to information I received while in New Zealand, one large squid species of the genus *Architeuthis* is occasionally seen in colder waters and is believed to grow at great depths in large numbers. Its fifteen-foot body has tentacles thirty to forty feet long with sucker disks as large as dinner plates. Each disk is capable of terrific suction power. In 1881 and 1888 three of these monster species (*Architeuthis longimanus*) were cast ashore near Wellington, New Zealand. One was fifty-seven feet long and was believed to be something near a world record for size. In 1959 a forty-seven-foot squid was measured at the Miami University Marine Laboratory.

Although seldom seen, it is known that the squid is vicious and will attack a human being as well as fight to defend itself against the huge sperm whale, which is its mortal enemy. Sperm whales are often captured with severe circular injuries and scars as evidence of battles with squid. A fifty-foot squid leaves a circular disk injury with a four-inch diameter. Injured whales have been found with eighteen-inch squid scars, four and a half times larger than those made by the largest specimens that have so far been seen. One can only speculate on the discoveries yet to be made in the great depths of the ocean. Little do we know the extent of the dangers that lurk beneath as we sail the peaceful-looking seas.

On October 26, 1876, two men and a boy were fishing near Portugal Cove, Newfoundland, when attacked by one of the monsters. It slipped a tentacle across the boat, pulling it toward its beak, which was double the size of a man's head. With all their lives in danger as the boat started to tip, the boy, Tom Piccot, only twelve years old, had the presence of mind to grab a hatchet from the boat and start chopping. They safely reached shore from their harrowing experience with a nineteen-foot tentacle as big around as a man's arm in their boat as proof that their nightmare had, indeed, been real.

On March 25, 1941, came further evidence of viciousness when the English troopship *Britannia* sank in the central Atlantic. Twelve men were clinging to a raft when a large squid tentacle pulled one to his death as the other eleven looked on in horror. Soon another man's leg was encircled by a tentacle, but for some reason he was not pulled under. When the squid released its grip, it left raw bleeding sores, and two years later the man was still carrying visible scars.

Flying squids of the genus *Illex*, also called sea arrows or arrow squids, are of a behavior class all their own. They travel in large schools and often leap eighteen feet clear of the water in some kind of frenzy to land on the decks of ocean vessels. They vigorously attack schools of fish with apparently a desire to satisfy some emotion, or craving, other than hunger, for they kill many more than they eat. Sighting a school of fish, the squid darts backward through the school with the speed of an arrow. With quick side turns, it grabs the panicked fish in the back of the neck. Using its parrotlike beak with great dexterity, it bites out triangular pieces of flesh in exactly the right way to sever the spinal cord, thus killing the fish instantly.

This arrow squid is oddly attracted by light. On bright moonlit nights many are stranded on shore. It is believed they become so fascinated by the moon that in swimming toward it they unknowingly enter shallow water and become stranded and helpless on the beach. Cod fishermen catch great quantities of this squid to use as bait. During my travels I have been told that local fishermen take advantage of this odd affinity for light to catch them. They simply build large bonfires on the shore at night. The squid oblige by stranding themselves in shallow water where all the fishermen need to do is pick them up. Each summer there is a mysterious mass migration of this *Illex* squid into the coves and bays of Newfoundland and Labrador. No young and no eggs are ever seen, and the life cycle is not fully understood.

The squid is considered a delicious food in many areas of the world. In fact, its flavor seems acceptable to all who try it—it is merely the thought that is frequently indigestible to many who fail in their attempt to eat it. In China it is an extra special delicacy and much is canned and sold to Latin and Mediterranean countries. It is also sold widely in the Chinatown areas of our own cities. It hangs in their shops looking so utterly repulsive, one leaves feeling a strong urge to become a vegetarian for the remainder of one's life. However, flavor seems to be winning out over repulsive appearance, for even in America the squid fishing industry is becoming important, especially in Cape Cod, Labrador, and Newfoundland. Squid is sold both for food and bait. In 1964 Newfoundland alone sold more than eleven thousand tons of squid to Portuguese fishermen.

It is a well-known fact that tremendous quantities of squid are used for human consumption, but getting comparative statistics is a difficult matter since, in the areas where they are consumed most heavily, people who eat them deny eating "fish bait."

Octopus

The familiar octopus, or devilfish, which belongs to the order Octopoda, is perhaps the best and most widely known of all the five cousins in the Cephalopoda group. Many people, however, are confused by the distinction between octopus and squid. It may help to remember that *octo* means "eight" and *pus* means "feet." Thus, octopus = eight feet (or rather, arms), whereas the squid has eight arms, plus two more extra-long ones, making ten in all, and can grow to a much larger size than its octopus relative.

The octopus is one of God's most widely scattered creatures. It can live near all types of shores—rocky, sandy, and muddy; in all temperature ranges—tropical, temperate, and polar; and in all depths, from shallow water to seas more than two and one-half miles deep, perhaps a greater depth range than can be survived by any other living creature.

"Beauty is in the eye of the beholder," and never were truer words spoken than with regard to the octopus. To some people the octopus is a thing of beauty. To the rest of us they are repulsive and frightening creatures. Although we all cheerfully grant each person the right to his own opinion, no amount of effort expended toward understanding will *ever* make one group understand the reaction of the other.

Even for those who are left shuddering with revulsion, these creatures hold a high degree of fascination. It is their ability to make instant color changes that holds one entranced. I have watched them crawl on the sea bottom beneath boats with their color continuously changing to match the color of whatever they were crawling over. One big brown one was just resting easily and quietly when, apparently for no reason at all, in a mere second, it became a beautiful brilliant red. When occasion demands octopuses can, for better concealment, produce different colors on opposite sides of the body through color-changing organs called chromatophores. Beneath the chromatophores lies a platelike layer of prism-shaped cells, called iridocytes, which catch and reflect the sun's rays, producing a shimmering iridescence that adds greatly to the fabulous color displays.

It is amazing, the speed at which octopuses can run on a dry surface. About a dozen partially grown ones escaped from the traps on a crab fisherman's boat where I was riding as a guest. The scattering of both animals and humans was as frantic as if a dozen big rats had made a dash from their cages. As human feet and octopuses' arms were dashing about, it would have been hard, indeed, for an observer to figure out just who was chasing whom. In water, however, the octopus will get from here to there by either crawling on its tentacles, or swimming by jet propulsion; that is, it fills its thick mantle with water and forcibly ejects it.

Some species are equipped with but single rows of suction cups, but usually each of the eight long arms, or tentacles, projecting from the bag-shaped octopus body has two rows of suction cups on the inner surface, and each and every one of those hundreds of cups has a very strong vacuum pull.

Thus, for those of us who feel repulsed, there seems ample justification, indeed, for our fear. Especially for me, since I have been an eyewitness to the damage those cups can do to a human.

The octopus is generally thought to be a timid creature wanting only to flee from a human being to a nice safe retreat. But I can vouch from personal experience that they do not always behave in so mild a manner.

On November 7, 1961, between Clearwater and Tampa, Florida, I was the only person near enough to help, so it fell my lot to come to the rescue of a swimmer attacked by an octopus. After that nerve-shattering experience, I hope all those who love (and see only beauty in the eye of an octopus) will forgive me for what I saw in the eye of that special octopus as it tried desperately to envelop me (as well as the swimmer) in its death embrace. Very close up indeed, I saw with complete clarity what, to me, was a cruel vindictive desire to kill.

Before that day it was only instinct that placed me in the ranks of the octopus nonlover. I had read of them capturing swimming dogs and I knew their bite could cause a very painful injury, but I had never read nor heard of an octopus attacking a human being in open water.

Since that day I have been told by others of similar octopus attacks on other people. Naturally, with that added knowledge, and with the look from those determined eyes seared permanently into my memory, this is one beholder who will be forever unable to see beauty, innocence and docility in an octopus, *especially* in its *eyes.*

The mouth of the octopus, located at the center of the circle of arms, is provided with a parrotlike beak. The animal crawls over the ocean bottom feeding chiefly on invertebrate animals. With its sharp beak, it nips a hole into the shell covering of its crab or mollusk prey. Then it injects a paralyzing saliva through the hole to liquefy the meat inside so that it can be sucked out through the hole.

The octopus also uses its arms in amazing ways to capture its food. The arms are so flexible they can pass a small captured animal from one suction cup to the next, the full length of the arm. They often grab into a group of small crabs, capturing a dozen or two at a time, and then leisurely pass them one by one up the suction-cup assembly line to the mouth to be eaten. In some species the very young have an even more amazing way of capturing their food. Until they grow strong enough to overpower their prey, they cut off strips of poison jellyfish tentacles the length of their arms. Then, holding the tentacles along the underside of the suction-cup-lined arms, they use them as a poison weapon to paralyze their prey.

The octopus not only forages over the sea bottom for food, but it often makes itself a den by hollowing out a crevicelike grotto. It then lies in its den with outstretched arms ready to capture any shell or fish that strays within reach. By changing its color to exactly that of the den, it becomes all but invisible to its prey. The doorway to its den is often deeply strewn with the bones and shells of its victims.

The octopus not only kills many animals, but is in turn killed by many. The eel is believed to be its most deadly enemy because, with its long slender body, it can probe back into the holes and crevices used by the octopus for refuge. However, a tremendous toll is also taken by dolphins, whales, fish, birds—and by man. To help the octopus meet its many enemies, nature plans well for its safety. It cannot only use color changes to make itself invisible when at rest, but by ejecting an inky fluid into the water, it can also make itself invisible in flight. The inky fluid not only conceals the fleeing octopus, but it is presumed to paralyze the enemy's sense of sight and smell. It can also confuse and temporarily halt a pursuing enemy with a brilliant flash of color. Nature even keeps its suction cups in prime condition so they can never wear out. Each day a thin layer of skin flakes off from each suction cup and is replaced by a new layer. And if it loses an arm to an enemy, another is quickly regenerated, just as with the lost arm of a starfish or the lost tail of a lizard. Clearly the octopus is extremely strong in its own environment, and yet, surprisingly, it is really a most sensitive creature. If placed in captivity, it often dies very quickly from nervous collapse.

A cold winter sends octopuses to deep water. Then, when late winter spawning time arrives, they come in great droves to shallow water. They are so terribly destructive to the lobster, crab, and fishing industries that fishermen use every means at their command to prevent them from spawning.

Some species of octopus lay several thousand tiny eggs no larger than a grain of rice. Members of other species lay less than fifty. The egg in my collection is a somewhat golden cream color, and appears to be very soft and fragile. Scientific knowledge has recently been greatly increased through a movie of a mother octopus caring for her eggs. The animal had chosen, or built, a hollowed-out cave little larger than her body. From the roof of the cave she attached many cords covered with thousands of eggs. The cords hung midway down in the cave like a huge mop. Throughout the incubation period the mother snuggled beneath and into the suspended mop, and kept it aerated by forcing a gentle stream of water through it with her siphon tube. And, very gently using her suction cups, she kept the egg mass cleaned of all foreign debris. A mother octopus eats no more food after her eggs are laid, and dies soon after they hatch.

From the moment the babies hatch, they are entirely on their own in an extremely hostile sea. Almost all the thousands that hatch are merely food for the schools of small fish assembled at the cave entrance waiting for them to emerge. Only a few survive that initial fish banquet. And, of those few, somehow, with the guiding hand of a Divine Providence, enough have escaped the multitude of other life threats to perpetuate the species on down through the centuries. Most octopuses die after spawning at about two years of age, but specimens with arms three feet long are estimated to be three or four years old. There are presumed to be about 150 different species, the largest of which, authorities say, has a body approximately the size of a football. Horror stories, cartoons, and movies have led many people to believe that the octopus is a real giant of a creature, so the football-size body will come as a surprise to some.

Octopus meat is considered a valuable seafood in many parts of the world, especially in the Mediterranean area, Japan, and Mexico. During the fall fishing season, as much as fifteen tons a day are captured to supply merely the Mexico City markets, and the demand still far exceeds the supply. When prepared by an expert, it is very tender and has an absolutely delicious flavor. I do not, however, personally recommend it—unless it *is* prepared by an expert. The one I prepared to satisfy my curiosity turned into a nice-looking plate of "rubber."

In Hawaii several species are caught and prepared to serve as *pupus*, which they eat for between-meal snacks.

At Puget Sound they have capitalized on their large octopuses to develop a most unusual sport. Each year a world's champion octopus-wrestling match is held. Thousands of spectators flock in to watch the champion team wrestle the largest octopus from the water to the shore.

The large Puget Sound octopus is reputed to have a gentle docile nature. But the divers manage to put on a most convincing fight with which to thrill the spectators. Possibly I may have been too easily convinced after my own personal encounter with one that was also reputed to be docile.

In Australian waters there is a small brown and blue octopus, barely four inches in diameter, which scientists believe is the deadliest creature in the world. Of the world's deadliest animals, the most liable to come to mind are the cobra, adder, and black widow spider. But compared to Australia's ringed octopus (*Hapalochlaena maculosa*), they are all sluggards at the kill. Their poison requires a matter of hours to do what the poison of the octopus can accomplish in minutes.

In June 1967 a twenty-three-year-old soldier, James Ward, was searching among the rock pools near Sydney, Australia. Attracted by its beautiful color, he picked up the tiny (less than four-inch) octopus, and in less than an hour he was dead. He did not know that when the ringed octopus is most attractive, it is most deadly.

Anger and fright deepen the ochre bands on its body and tentacles, and change the dull blue rings to an iridescent peacock blue, making it a highly desirable trophy for the unwary collector. When in that state of color change, it can quickly inject enough venom to kill seven men, a venom that acts so quickly that even if an antidote is ever developed, it is unlikely that it could be administered in time.

Within five to ten minutes of being bitten, the victim starts acting as if he were drunk. Muscles around the face and neck become weak and numb, and breathing becomes difficult—then soon impossible. The only possible help for the victim is to immediately be given oxygen and artificial respiration. One case is on record where a schoolteacher's life was saved in that way.

The Commonwealth Serum Laboratories of Australia have researched the ringed octopus and they pronounce it the fastest killer of any creature on land or sea.

The octopus is now playing a special role in modern medical research. A substance called cephalotoxin has been extracted from the octopus saliva. It is a complex poison with the power to hinder blood coagulation. Another saliva extract called elecoisin has the ability to control high blood pressure and to correct the irregular heartbeat often associated with heart seizure. It is believed that with further experiment these extracts may in time fill an important need in the world of medicine.

Medical researchers are also finding the brain of the octopus very helpful in studying human brain disorders. The octopus brain is physically well developed—the octopus is believed to be superior in intelligence to any other sea animal except mammals, and is capable of learning—but is still uncomplicated enough to be useful in a variety of experiments. Among other things, researchers are using the octopus to find out precisely how a brain performs when influenced by drugs. Scientists have also found that the octopus reacts to certain situations with apparent emotional responses, quite like the human emotions of rage and fear. The octopus also has well-developed eyes with very keen vision.

MOLLUSKS AND MAN

14. Food from the Sea

Back through the ages, man has looked to the animals of the ocean for an important part of his diet, and to the plants as well. Many mineral-rich seaweeds, kelps, mosses, and algae are highly nourishing and now, with more appetizing methods of preparation, they are gaining more widespread acceptance around the world. Seaweed has long been a staple of Japanese cookery, and the Chinese too have put these foods to good use. One Chinese seaweed-based delicacy, bird's-nest soup, is truly a triumph of ingenuity. The Chinese wait for birds to line caves with nests built from the seaweeds they want, and then very simply and easily collect the nests for their soup, a laborsaving device if there ever was one! Most Americans would turn up their noses at such an idea, but, in fact, we now consume with relish tons of a similar seaweed extract, called agar-agar, in our own soups, packaged desserts, ice creams, cakes, and baby foods. The only difference is that we laboriously collect our own weeds, while the Chinese let the birds do it for them.

Seaweed may sound like an alien food, but everyone is familiar with some among the hundreds of species of edible ocean fish, and most people have at least a nodding acquaintance with crabs, lobster, prawns, crawfish, shrimp, and, more rarely, sea turtles. But most Americans, at any rate, are surprised to learn that squid, octopus, eel, sea urchins, and the ugly but plentiful sea cucumber are popular staples in many parts of the world.

Mollusks, too, have played a tremendously important role in man's diet since he and his insatiable appetite first arrived on the scene. Archaeological excavations in many areas of the world reveal kitchen middens filled with huge piles of shells.

The seafood mollusks that come most readily to mind, and which we consume by the thousands of tons all over the United States, are the common oysters and scallops, as well as, in these days of refrigerated shipping, clams, mussels, and abalones. Less well known in this country are cockle shells, jingle shells, and the plentiful little limpet shells that cling to rocks in almost every area of the world. In many areas the limpet is eaten only as a last resort against starvation, while, ironically, in the Hawaiian Islands it is served to tourists (at extremely high prices) under the name opihi. In the South Pacific the exotic-sounding tiger cowrie is a major food. Tons of them are tossed on hot coals and roasted, or else the meat is dried and fried in oil. Another little-known food mollusk is the odd little coon oyster—a boon to hungry fishermen. It grows conveniently attached to a root or branch of the saltwater mangrove plant, and all the fisherman need do is break off the branch and toast it as he would a marshmallow. Ark shells also serve many food needs. One species, *Arca granosa,* has vivid red flesh that is believed by the natives of the Madras coast of India to be especially strengthening. Pregnant women eat quantities of the meat to help prepare for childbirth. Actually, there are many more edible mollusks that sound strange to our ears. A few tasty examples are the *Venus, Crepidula, Crassostrea, Mytilus, Polinices, Busycon, Ensis, Mya, Pinna,* and *Littorina.*

In some species, such as the abalone, only the foot muscle is eaten. In others, like the scallop (*Pecten*), the adductor muscle, which draws the two shells of a bivalve together, is the only edible portion. There are still other species—the common food oyster, for example—in which the entire animal is consumed. The very small species, like *Littorina,* are prepared right in the shell, and the cooked meat is extracted with a toothpick.

But for now, our concern is with the common food oyster, a very small mollusk with the very large name *Crassostrea virginica* Gmelin, popularly called the Chesapeake Bay oyster. So popular has it become, that an amazing industry has grown up around it. I hope a little probing into the workings of that industry will prove of interest to those who may like to know just how that oyster managed to get into their bowl of soup.

Notes on Oysters

The ugly little food oyster with the jawbreaker name inhabits the whole of our Atlantic Coast. And, by carefully packing its seed in damp seaweed, it has been flown alive three thousand miles to the Pacific Coast, where it is now thriving. Look at that one oyster on your spoon, and then imagine enough of them on a stack to fill five million bushel baskets. Just from Chesapeake Bay alone, that many are harvested and eaten each year.

A number of years ago, the United States industry began with a wasteful exploitation of the extensive natural beds of the Chesapeake Bay and Long Island Sound. In the nick of time, before the natural beds had been completely destroyed, scientists researched the life history of the animal, and the information was used to increase production by underwater oyster farming.

The research revealed that, in early

One of the author's shell designs. *Robert Rickert*

Plate 26

(a) A lovely example of the *Murex ramosus* Linné shell (8–12 inches long).

(b) A venus comb murex shell (6 inches long), one of the most lovely of the Muricidae family, showing its long graceful spines.

(c) A 3-inch-long olive shell (*Oliva porphyria* Linné) found in Panama. No two shells ever have exactly the same pattern.

(d) A live *Oliva sayana* Ravenel (2½ inches long) burrowing through the sand, leaving a molelike trail in its wake.

(a) A 3–4-inch lion's paw (*Lyropecten nodosus* Linné) dredged from deep water in the Gulf of Mexico.

(b) A spectacular view into the gaping valves of a live 2½-inch bay scallop (*Argopecten irradians*), one of the common food scallops.

(c) Rigid pen shell (*Atrina rigida*)—8 inches.

(d) An extremely rare *Pleurotomaria* slit shell (3–5 inches across) found in the South China Sea off Taiwan.

Plate 28

(a) A live moon shell (*Polinices duplicatus*). The striped bulb-shaped portion in the center is the shell (2 inches across), resting upon the flat pancake shape of the animal (3½ inches). The two pointed spikes on the right are its tentacles.

(b) The lovely shell of the spider conch (*Lambis digitata*)—5 inches across.

(c) A 10-inch queen conch (*Strombus gigas* Linné) shell.

The spiny oyster (*Spondylus americanus* Hermann), one of the most beautiful American mollusks—6 inches across.

Plate 30

(a) The sunrise shell (*Tellina radiata* Linné)—2 inches long.

(b) The Florida dye shell (*Thaïs haemastoma floridana* Conrad)—2 inches long.

(c) The grinning tun (*Tonna ringens* Swainson)—4 inches long.

(d) Paper fig (*Ficus subintermedius*)—3 inches long.

(e) The great screw shell (*Turritella terebra* Linné)—4½ inches long—which inspired Archimedes with the idea for the common screw.

Plate 31

(a) Triton trumpet (*Charonia tritonis* Linné) found in the Pacific—14 inches long.

(b) The great top shell (*Trochus niloticus* Linné)—5 inches high. On the left the overall top shape is clearly defined, while in the right-hand picture you can see the aperture located in the bottom of the shell.

(c) An interesting display of the big green turbo (*Turbo marmoratus* Linné)—also known as the giant green snail—prepared by Florence Kuczynski for the Saint Petersburg, Florida, shell show. The green shell in the center (8 inches) is the snail in its natural state, with its white operculum on the right. The surrounding objects show how the shell appears polished and carved.

Plate 32

(a)

From the author's collection—*Left:* (a) *Voluta cottonia.* Caught from an Australian fishing boat, this shell is ordinarily the same size as the *roadnightae,* which makes this outsize 14.4-inch-long specimen of unusual value. *Right: Voluta roadnightae,* about 5 inches long. *Center bottom: Voluta bednalli,* approximately 4 inches long. *Robert Rickert*

(b)

(c)

(d)

(e)

(f)

A series of volute species. (b) *Voluta ebraea* (6 inches long); (c) the Florida *Voluta junonia* (5¼ inches long); (d) *Voluta bednalli* (4 inches long) of Northwest Australia; (e) *Voluta magnifica* (8 inches long) from eastern Australia; and the (f) *Voluta imperialis* (7 inches long), which inhabits the Philippine Islands.

Precious wentletrap (*Epitonium scalare* Linné)—2 inches high.

Plate 34

(a) Florida's pear whelk (*Busycon spiratum*)—4 inches long.

(b) The author's matched pair of 4-inch left-handed whelks. *Left:* this shell, with its opening and thin lip bulging out to the right, is the unusual "right-handed" left-handed whelk. *Right:* a typical left-handed whelk. *Robert Rickert*

(c) A group shot of the chambered nautilus, with shells ranging from 4 to 8 inches.

(d) Live paper nautilus. *Peter L. Vila, Courtesy Wometco Miami Seaquarium.*

(e) One-inch "ram's horn" shells taken from within the tubelike body of the *Spirula spirula* animal.

Left: (a) a pearl farm in the Inland Sea of Japan, showing the oyster rack as it rides the water. (b) *Right:* the oyster rack out of the water. *Charles E. Herron*

(c) Pearl oysters being seeded with small bits of clamshell. *Charles E. Herron*

(d) Sorting pearls. *Charles E. Herron*

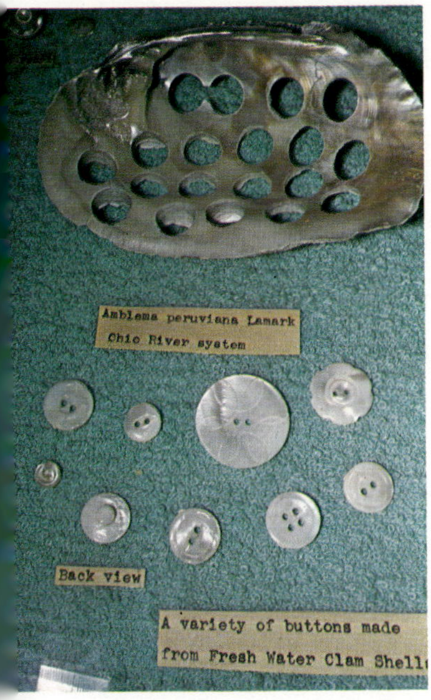

(f) Pearl buttons from river clams, and a clamshell with the buttons stamped out.

(e) A 4-inch quahog clam with pearl.

Plate 36

(a)

(b)

(c)

(d)

(a), (b), (c), (d) A few possible sites for locating live mollusks: *Top left:* small shells nestling in the roots of the mangrove. *Top right*: horse conch clinging to a wall. *Bottom left:* chiton underneath a rock. *Bottom right*: angel's wing siphon poking up through the sand. (This shell, which is usually 1 to 2 feet under the surface, should preferably be carefully dug up by hand to minimize the chance of shell breakage, rather than grasped by the siphon.)

summer, the oyster gill chambers become gorged with a milky fluid: in the female, the ova, and in the male, the milt, or sperm. When ripe, the ova and milt are simultaneously discharged into the water and their union (the fertilization of the ova) occurs. This, by the way, is another intriguing example of the biorhythm mentioned in Chapter 2. In the oyster, as in many other species, the clocks of both male and female are set at the same minute. If the clock of one was even slightly behind schedule, the cells of the other would dissipate in the moving water, the union of the two would be impossible, and the species would become extinct. But somehow it works out all right, and the water soon swarms with fertile eggs. It is estimated that a large oyster can produce as many as sixty million eggs. If every descendant of a single oyster lived, in five generations they would fill a space six times the size of the earth.

The whirling millions of young embryos (or fry) give the water a cloudy appearance. During this critical time, the oyster grower is nervous with apprehension. In a day or two the free-swimming youngsters settle to the bottom, and favorable weather at just this time is vital to a good set. A cold spell can kill millions, and a rough sea drifts millions out into deep water where they perish.

Once the spat—the baby oyster—has settled, it chooses, as if by instinct, a rough limy surface, which feeds it the constant supply of lime in solution it requires for continuing growth. A disintegrating dead shell is a favorite place for attachment. With the lime, oxygen, and minute plants and animals in the water to nourish it, the oyster, which is fixed for life, grows about an inch each year, having assumed the characteristic form of the adult when no larger than a pinhead. It continues to lengthen for about five years, and then it grows in thickness. The average oyster's natural life is about ten years. In ideal circumstances, undisturbed oysters have grown to a size and thickness that warrant the estimate of an extreme age limit of fifty years. The meat from the largest oyster that I ever found, when hulled out, covered my entire hand, and when cooked, had to be cut up in bites like a serving of steak. Its shell is now in my collection—of course, highly treasured, although very ugly.

The natural life cycle sounds simple—it is easy enough, surely, to set up the proper conditions to grow oysters on a commercial scale. Just throw around a few dead shells for the spat to sit on, and let nature take its course. But in fact, the problems of establishing an artificial underwater farm, or bed, are monumental. First, the grower has to keep in mind that oysters both breathe and eat from the same current of water. The oyster must bathe its gills in the moving water to keep its blood oxygenated, and it strains out small particles of food from the flow to keep itself nourished. Twenty gallons of water must flow through the body of each oyster each day to satisfy its needs. The temperature of that water must be just right, and it must have just the right levels of oxygen and sufficient food content or the oysters will die. The proper locations, following nature's pattern, are free from a strong current, supplied with food, and range in depth from fifteen to thirty-five feet.

A number of methods are used in making the beds. For example, in China the spat are grown on bamboo screens set in river estuaries. The French use hollow tiles, coated with cement inside and out, which are piled on trays suspended between posts under water. The most usual method in this country is flat-bottom farming, where the grower merely clears away the debris and covers the ocean bottom with old shells for the spat to settle on (Figure 12).

If nature is kind to the growing oysters and their grower, and the set is good, the spat must be thinned to prevent overcrowding when they are a year old, and about the size of a quarter. With the surplus the grower can extend his own acreage or he can sell his seed to other oyster growers.

But the odds against smooth sailing are huge. It has been estimated that any given larva has but one chance in two million of growing up. Fish devour the larvae in great quantities, and even after they have settled and grown a protective shell, they must still face a multitude of dangers.

A freshet may cause smothering deposits of mud, or the currents may shift the sand enough to bury them. Crabs many times devastate an entire bed, crushing all shells less than a year old in their powerful pincers. Drills (*Urosalpinx*), dog whelks (*Nassarius*), the moon shell (*Polinices*), and the pear conch (*Busycon*) kill millions by boring through the shell with their rasping tongues and sucking out the soft oyster meat.

One of the most insidious enemies is the boring sponge. It honeycombs the shell, and the oyster becomes exhausted sealing up the punctures with new lime deposits.

The small (less than an inch) oyster borer (*Lepsiella scobina*) causes great destruction and expense. It can bore through the hardest oyster shell with the aid of a sulfuric acid secretion in less than half an hour. This acid causes a chemical change in the shell that softens it somewhat.

Starfish, ray, and octopus take a tremendous toll. The starfish begin their destruction when the oyster is scarcely the size of a pinhead, and they continue throughout its life. They wrap their arms about the shell and hold it until the victim's strength is gone. Then they push their inverted stomachs between the valves of the oyster and digest the living meat. Starfish go in swarms, migrating from place to place. Many times they completely destroy an oyster bed before the owner even suspects anything is wrong. To fight the problem, a great mop, or tangle, made from bunches of raveled rope has been devised. This tangle is dragged over the beds to entangle the starfish so they can be lifted to the surface and destroyed.

Thick-growing algae often smother oysters. The farmers have learned to control it by scattering large quantities of algae-eating periwinkle shells over the beds.

If he is to succeed, the oyster farmer must relentlessly fight nature every inch of the way, just as the corn or hog raiser must.

When the farmer, by periodic checks, feels his oyster crop has arrived at prime marketing condition, the harvesting procedure begins. The first and most primitive harvesting method entails going out in small boats with tools called nippers, and lifting the oysters up one by one. In the second method dredges drag over the bottom, and then the shells are picked out from the rubbish that the dredges bring up.

With the invention of a new harvesting machine, a third technique has recently come into operation. Using a high pressure water jet system, the machine brings the shells up by an airlift. With this system, two men can harvest a hundred bushels of oysters in ten minutes from water eight to sixteen feet deep.

It is a fallacy, by the way, that oysters are not edible during the summer months that do not contain the letter *r*. However, they are never harvested then because this is their spawning time, and while they are spawning they lose fat and are less tasty.

Once harvested, the oysters must then be sorted, all by hand. The top quality are separated by size—large, medium, and small. The poorly shaped are kept apart to open and sell as bulk oysters. The undersized are returned at once to the water for further growth, and the dead are used to help line fresh oyster beds.

The sorted oysters are ready then for the final touches that, giving nature a boost, bring them up to a perfection of flavor and size fit to grace any dining table. First, oyster farmers have found, after much experimenting, that its flavor can be much improved by fattening it for a time in a tank well supplied with finely ground cornmeal. Then the animals are put through an unusual procedure called plumping. For a day or so, they are placed on perforated rafts with shallow compartments. The rafts are placed in canals that are cut in from the beach, but receive freshwater drainage from the land side. In this water, which is much less dense than seawater, the oysters bloat and seem to take on more flesh. The gills and alimentary tract are cleaned, which is especially desirable when they are taken from muddy beds. In actual nutrient value, oysters lose about 13 percent, since they gain from 12 to 20 percent additional weight by reason of the water they absorb. Left too long on the rafts, they become lean and tough and lose flavor. So the plumping process is very carefully timed. Extreme care must be taken to guard against contamination of the canals, or the oysters will become carriers of disease.

Oysters, just as all other filter-feeding seafoods, pose a constant bacterial contamination problem. Indeed, from the beginning of recorded history, mollusk bacterial contamination has been something to reckon with. Historians suspect that, far back in the fourteenth century B.C., Joshua captured the city of Jericho not as a military victory, but because the defenders of Jericho were too ill from mollusk bacterial infection to fight. With modern twentieth-century water pollution growing steadily worse, the problem with mollusks increases proportionately. In the past the most common method of treating suspect mollusks has been to place them for a time in chlorinated water. The method is tedious and often leaves an objectionable chlorine odor on the meat, and thus prompted further research. A Japanese researcher, E. Tanikawa, at Hokkaido University, has now devised a new method that is both more efficient and more effective. He has learned that when an electric current is applied just right, the bacteria within the meat (even the more dangerous ones that resist the heat of cooking) will migrate to an electrode where they can be removed.

With the oyster nicely fattened and plumped, the farmer has at long last completed the difficult task of producing an oyster ready for market. The shuckers and packers then take over and do a most efficient job hastening it on its way to your grocer. The shuckers kill the animal by breaking the thin bill from the shell with a hammer. They then skillfully cut the muscle that holds the shells together, using a special oyster knife to open the shell and remove its contents. Each shucker receives his tally check when he empties his bucket into the conveyor trough. The shucked oysters are then carried through a perforated strainer to drain away the liquid. Bits of shell are removed as the oysters move along to be measured and thoroughly washed (a few gallons at a time) in a colander.

They are then packed into quart cans, kegs, barrels, and tubs, stored in refrigerated cars, along with cakes of ice and sawdust, and successfully shipped inland for long distances.

Happily for all concerned, a wholly new method of oyster culture is now showing great promise. It is believed that by this revolutionary method it will be possible to produce sixty-four thousand pounds of oysters per acre—several times the present harvest by conventional methods. This new technique is also expected to open up new areas for oyster culture and thus multiply man's bountiful harvest from the sea, although it is more costly, more difficult, and requires more skilled help than the traditional method.

This new procedure calls for placing the oyster seed on a hanging clutch, and letting it grow there. The seed (larva) is planted on shells, about thirty to the shell, and the shells are spaced and attached thirteen to a string. Wooden racks are then constructed in water deep enough to suspend the strings so the bottom shell does not touch bottom.

Thus suspended above its many bottom-feeding predator enemies, the oyster has a more than 80 percent chance of survival, as contrasted with the 20 or 30 percent chance by the traditional mud-bottom method. Also, thus suspended, the oyster pumps the plankton-filled water nonstop, thereby accelerating its growth until it is ready for harvest in half the time—about a year and a half as compared to three years by the old method. And, a final advantage, the harvesting procedure is greatly simplified, which counteracts the more costly planting operation. The strings are simply lifted from the water to retrieve the crop.

The experiment is being carried out by the Drake's Bay Johnson Oyster Company in Drake's Bay, forty-five miles north of San Francisco. The plankton-filled waters of cool Drake's Bay are ideal for oyster growth, but the temperature is too low for adequate production of the seed, or larva. Up until now the company has imported most of the twelve million seeds they have used from Japan. But the seed problem is now well on the way to being solved by California's San Mateo Pigeon Point Laboratory. At long last techniques have been perfected to produce oyster seed in commercial quantities.

The United States Bureau of Commercial Fisheries is now alerted to the need for urgent action to revolutionize the oyster industry. During the last sixty-five years the population of the United States has more than doubled, while our oyster supply has shrunk to less than half. Clearly that trend must be reversed if future generations are to be supplied.

For thirty years our oyster industry declined from pollution and backward management, while other countries such as Spain, France, Korea, Australia, and Japan have increased their production by the newer off-bottom method. Their success has been so great that Japan, alone, now annually exports thousands of bushels of seed oysters, and Spain is even more advanced.

It is believed the oyster production potential for our Chesapeake Bay region is as great as that of Japan or Spain. In 1960 the Bureau of Commercial Fisheries Biological Laboratory at Oxford, Maryland, began an extensive study of the off-bottom method of oyster culture. They have since conducted many experiments in an attempt to determine proper location for and proper timing of each operation. Progress is hampered by the predator flat worm *Stylochus ellipticus* that is prevalent in the Chesapeake Bay area. Though the off-bottom method places the young oyster out of reach of all bottom-crawling predators, this flat worm has a free-swimming larva that sets on the suspended shells and feeds on the oyster spat, inflicting great damage. Research is now being undertaken to control the worm.

It staggers the imagination as one tries to predict the far-reaching impact on the future oyster industry of the world if, or when, the new methods of seed and oyster production come into wide use on both our East and West coasts. With the present rate of progress, the United States may eventually take its proper place in the oyster industry of the world.

Figure 12. The shell planting process in operation. *Bob Williams, courtesy U.S. Fish and Wildlife Service*

15. Ocean Jewels

Few people have stopped to wonder how lustrous pearls of a necklace, or our familiar pearl buttons, or the lovely mother-of-pearl used for jewelry and exquisite inlaid work, could grow beneath the waters of the ocean or, for that matter, in a muddy river bottom. And, even those who do wonder are confused by the different terms. Just what is *pearl,* as in a pearl button? Is it the same thing as mother-of-pearl? If not, how do they differ? What is meant by a pearl shell or a pearl-lined shell? Are they the same thing? And last but certainly not least, how does the lovely round jewel that we call pearl grow, and what is the difference between a natural pearl and a cultured pearl?

As we clarify these terms, we shall explore some of the most intriguing processes in the sea. The factories in which our gems are manufactured are the bodies of three altogether different types of mollusk. The mantles of these three types, just like every other mollusk, secrete a fluid substance that hardens to form the animal's shell and to provide a smooth lining for the inside of the shell. The difference is that, in most species, the fluid is dull and lifeless, and thus hardens into dull lifeless shell and lining; in a few species, however, the fluid is pearly and lustrous in quality, and it is this lustrous secretion—called nacre—that forms the raw material for jewel pearls and for the substances we know as pearl and mother-of-pearl.

Pearls

The Egyptians are believed to be the first people to ornament themselves with the rare and valuable jewel pearl. They believed that a god discovered them off the coast of India some four thousand years ago, and Egyptian women were known to have worn pearls from about 1500 B.C.

The famous, beautifully shaped oriental pearls come from the Persian Gulf, Ceylon, the Red Sea, and the Gulf of Aden near Zanzibar, East Africa. They were popularly used as adornment from the time of Aristotle and Alexander. Prominent women among the Medes and Persians would wear two or three pearls strung on a ring through their nostrils.

The people of ancient India and China had great faith in their medicinal value. They ground pearls to a fine powder to cure lunacy, leprosy, and other ills. The Chinese also, along with the early Hindus, Japanese, and the Indian mound builders of the Mississippi Valley, placed pearls in the mouths of their dead. Pearls in enormous quantities were found in the temple of Tolomecco in Florida, and Montezuma's temple in Mexico was heavily decorated with them. Indeed, the Indian chief's main source of wealth came from pearl fisheries.

The Romans used pearls very extensively to decorate their temples, chariots, clothing, and furniture. History tells us that the pearls for which Britain was noted spurred Caesar to cross from Gaul to conquer Britain. He returned with a highly coveted breastplate adorned with pearls, which he dedicated to the Roman Temple of Victory. England stayed especially proud of her pearls. Much later, during the reign of Queen Elizabeth I, in an attempt to impress a visiting Spanish ambassador, the wealthy merchant Sir Thomas Gresham drank a wine toast to the queen, in which he had dissolved a famous pearl worth £15,000.

The most perfect known pearl, La Pellegrina, in Moscow's Zosima Museum, is a twenty-eight-carat globular Indian pearl of unusual beauty. The largest known pearl, in the Victoria and Albert Museum in London, is two inches long, four and one-half inches around, weighs three ounces, and is somewhat irregular in shape. In 1883 a famous pearl of great value was found in western Australia. Nine pearls had grown together in the form of a one-and-a-half-inch cross. Known as the Southern Cross, it was sold for several thousand pounds. However, its ultimate fate is uncertain. Another, and possibly the finest Australian pearl ever found, is the Star of the West. Valued at £14,000, it, also, was found on the northwest coast (in 1917), weighed one hundred grams, was pear shaped, and the size of a sparrow's egg.

For centuries men were constantly lured into great danger—and often lost their lives—in a never-ending quest for this highly coveted and extremely valuable jewel. And all the while they searched in vain trying to unravel the elusive secret of just how and why certain shells produced valuable pearls, while others, even of the same species, did not. Many legends grew up about the origin of pearls. One curious notion had it that pearls were formed by raindrops falling into the gaping valves of the pearl shell. Others believed that little seed pearls would reproduce if closely confined and covered with rice.

Legends finally lost their force, however, and ancient records show that Professor Louis Boutan of France made the discovery that pearls formed when a bit of foreign matter became lodged within the shell of a mollusk. If unable to dislodge the irritant, the animal would, for greater comfort, coat it with the same gland secretion with which it lined its shell. Purely as a matter of self-protection against the irritant, the glands would secrete an extra amount of the fluid, just as our tear glands increase activity if a speck lands in the eye. Only, with the mollusk, the offending speck, instead of being washed away, is coated over and over with the liquid gland secretion, with each layer hardening to eventually become the smooth, round nonirritating object we call a pearl.

In nature the production of a pearl is an extremely rare occurrence. Only about one mollusk in each thousand produces one and, of those produced, only about 5 percent are of sufficient quality to have commercial value. So, once the basic principle was grasped, it is understandable that people began groping toward the possibility of pearl culture. But, for centuries no headway could be made because nature concealed her secret so well it could not be fully comprehended.

Later records indicate that a thirteenth-century Chinese prince is credited with the first faltering progress in artificial pearl culture. He observed that some mollusks produced blister pearls, a kind of half pearl still attached to the shell, whereas others contained round valuable pearls free within the body cavity. He drew the mistaken conclusion that the blister pearls gradually grew to a round shape and detached themselves. He then introduced foreign objects between the mantle of the mollusk and its shell, hoping to stimulate growth of blister pearls that would later detach, but his experiment was only partially successful. He did produce the blister pearls, but nothing more.

It took centuries more study and experimenting before the key to nature's secret was learned well enough for man to aid her in more rapid production of the valuable round pearl. Not until the beginning of the twentieth century was real progress made. At that time two discoveries of vital importance were made. It was learned that, to produce a cultured pearl by artificial tampering with nature's processes, the foreign object had to first be coated with living nacre-producing cells before placing it in the body cavity of the mollusk. This was done by snipping off a small piece of the living mantle containing the fluid-secreting glands. This small mantle snippet was then placed, along with the foreign object, into the body cavity, and its secreting cells would cause a cyst, or pearl sac, to be formed around the foreign object. Then, inside the cyst, the cells would continue their normal function of nacreous secretion to gradually add layer upon layer around the object until a smooth round pearl was formed.

The second vital discovery was that there was not merely one but six different zones in the body cavity in which a cyst would develop. With still more experimenting, to learn which zones would yield what results, they were at long last almost ready to begin actual experiments with pearl farming, or the raising of cultured pearls.

But first there was one more major problem to solve. Which mollusk species would adapt best to pearl farming? Beautiful pearls had been found growing naturally in the *Strombus*, *Trochus*, abalone,

Pearly Nautilus, *Mytilus edulis*, and many other mollusks. But, because of its exceptionally lustrous nacre, the small marine *Pinctada martensi* species, which grows in tropical waters, has an outstandingly beautiful pearl-lined shell and was selected as the mollusk best adapted for the purpose. The *Pinctada martensi* oyster owes its prominence to K. Mikimoto, a Japanese pearl farmer, who is responsible for the greatest contribution to the development of pearl culture in the twentieth century. It was he who, with several generations of careful breeding, developed the *Pinctada martensi* into a strong superior mollusk for pearl culture.

With a name like that, it is understandable that the *Pinctada martensi* soon acquired an easy-to-pronounce nickname—the pearl oyster.

That little word *oyster*, by the way, simple though it is to say, has caused much confusion over the years. So many different mollusks, of so many different families, are popularly called oysters, that the question of just what an oyster is anyway seems to be a big puzzle indeed. Actually, however, there is no oyster species or family. That is just a simple nickname applied to many different kinds of mollusks, because their real names are too hard to pronounce.

For example, the ugly, heavy-shelled, oblong *Crassostrea* oyster, just discussed in our food chapter, is an entirely different animal than the smooth, fragile, roundish *Pinctada martensi* jewel oyster of this chapter—and neither of them is anything like the beautiful big *Spondylus* oyster with its fantastic long spines, or the ugly little *Ostrea frons* coon oyster, or any of the other so-called oysters.

Pearl Farming

So many thousands of pearl farms have now gone into operation in many areas of the world that pearl culture has now assumed the proportions of a major world industry. The farms of Mr. Mikimoto, who perfected the method of coating the foreign object with living cells, are the most famous in the world. Indeed, Japan is now the major source of cultured pearls. In the small coves of Argo Bay, about 190 miles southwest of Tokyo, there are now over three thousand pearl farms. In 1967 Japan alone produced 37.7 tons. Their 4,200 producers, 500 processors, and 170 exporters worked in strong competition for the foreign markets. The United States alone bought over $16 million worth of pearls from Japanese sources.

The actual procedures involved in the production of cultured pearls make a fascinating tale. For nowhere in the world does man struggle so hard against the forces of nature for financial gain.

Using the improved strain of the *Pinctada martensi* developed by Mr. Mikimoto, the spawn is raised in special underwater cages (Plate 35 a and b). Then, when the oysters are two years old, they are transferred to other underwater cages for another year. At the age of three, mature, and with a life expectancy of another ten years, they are ready for the inoculation to produce culture pearls.

An operator, usually a nimble-fingered girl, now begins her task by cutting tiny snippets from an oyster's mantle (the membrane that produces the iridescent pearl nacre). She then places into a clamp a live three-year-old oyster whose shell is already open about one-half inch after being removed from the water. Peering into this opening, which cannot be widened without killing the oyster, the operator carefully cuts into the living stomach with a miniature scalpel and inserts two of the tiny mantle snippets into the incision. She then implants two tiny fragments of clamshell so that each touches the small pieces of mantle tissue (Plate 35c). Then, with the hope that the oyster may produce two pearls, it is placed back into the bay to live in a wire basket suspended from a raft.

The transplanted mantle soon envelops the clamshell nucleus and coats it with a layer of nacre, just as it might line the inside of a shell with the pearl substance. From three to seven years later, if all goes well, the oyster will yield two perfect pearls. But, unfortunately, the chance is astronomically high that the oysters will not produce the perfect cultured pearls hoped for.

To fight his way to success, the hard-laboring pearl farmer must cope with the devastating effects of typhoons, smothering growths of barnacles, moss, algae, not to mention red tides, disease, and a number of other destructive forces, including extreme temperature changes. The changes of water temperature, due to changing ocean currents, wreak havoc if not dealt with properly. In one method of protection from killing temperature changes, the cages suspended from rafts are raised or lowered to correct temperatures; another way is to move the rafts by boat to a warmer or cooler area.

Until recently, huge losses resulted from infections in the scalpel incisions made to insert the tiny seed pearl. But now the seeds are first dipped in the antibiotic solution Aureomycin Chlortetracycline. Its use has resulted in a 30 percent increase in top quality pearls. Indeed, the quality of the cultured pearl has now become so good that only an expert can distinguish it from the natural. And even he, in most cases, cannot be sure from an external inspection. The only sure way is to X ray the pearl, subject it to a powerful electromagnet, or drill into it. In some cases all three methods must be used.

When the pearls are harvested they must be hand sorted for size and color (Plate 35d). The laboratory recognizes six main color divisions: white, cream, pink, green, gold, and black. Assembling matched pearls for necklaces is carried out with great attention to color as well as to size and shape. Pearls that match are lined up in velvet grooves before being strung on silk threads. The sorting is so delicate and demanding that the finest expert cannot assemble more than half a dozen first-class strands in one day.

The Japanese have now developed a new way to make captivating pearl novelties of great interest and value. Instead of inserting a small clamshell seed as a nucleus, they insert tiny Buddhas, tiny crabs, or other tiny sea animals and objects of shapely interest. The oyster is then returned to the sea and given time to coat the object with a layer of nacre, but not time enough to build the nacre to a thickness that would conceal the shape of the object. The result is a very lovely little pearl jewel object.

In my collection is a very unusual pearl freak that is purely a product of nature.

The factory of Mr. Umlandt and Mr. Worst had been in operation since 1898. At the time of my visit, it was producing 7,200,000 buttons a week, employing 150 persons, and using 2,000 tons of clams per year. Buttons were shipped to every section of the United States, including my own home town of Huntington, Indiana. It so happened that Mr. Umlandt recognized, as his own, the pearl buttons on the dress I was wearing.

Originally, the company had gotten all their clams from the Mississippi River, but at the time I was there dam construction, overcollecting, and water pollution from large cities had largely destroyed clam growth in the Mississippi area, so they were buying their supplies from the Tennessee River.

In more recent years overharvesting has also depleted the supply in the Tennessee River. Although modern synthetics have lessened the demand for pearl buttons, the rapidly growing Japanese pearl culture demand began by 1960, to exhaust the supply in Tennessee and Kentucky, and the industry started moving northward into Indiana. The Wabash River system is now one of the outstanding clamming areas of the United States.

Recognizing the importance of the industry, and the need for regulation to prevent depletion of supply, the Department of Natural Resources is coming forth with timely action. A closed season is now enforced from November 1 to April 15, and the University of Louisville is researching the project to determine the extent of fishing to permit each year. However, before it will be possible to restore the Wabash River mussel population to its original abundance, a way will have to be found to divert the poisonous industrial wastes that are now even more destructive than overcollecting.

Mother-of-Pearl

There is still another shell of special fascination that combines qualities of both the small, fragile, lustrous thin-shelled jewel pearl oyster and the thick-shelled, pearly river mussel, yet is like neither of them: the mother-of-pearl shell. It is far larger than either, often growing larger than a dinner plate. Its thickness, often a half inch or more, even exceeds that of the river clam. But the great distinction of the mother-of-pearl shell is that it combines a superior, brilliant, iridescent nacre equal in quality to the nacre of the jewel pearl oyster, with a whole thick shell constructed with layer upon layer of the same lustrous substance. Just like the river clamshell, the mother-of-pearl shell adapts beautifully to the making of buttons, jewelry, inlay work, and miscellaneous ornaments, with one important difference: It is so superior in quality. If you will compare your pearl (that is, river clam) button with your mother-of-pearl button, you can easily distinguish the difference.

Though a number of mollusk species build up their shells with the lovely mother-of-pearl nacre, in this chapter we will discuss only the *Pinctada margaritifera*—the great pearl oyster—which is the major commercial source of the world's mother-of-pearl. The great pearl oyster will also occasionally produce a valuable jewel pearl, but it does not adapt well for use in the cultured pearl industry.

These oysters grow in most tropical seas and their nacre, which varies greatly in color, is most often white, golden yellow, or dark smoky gray. They anchor themselves to a rock or any solid surface in the bottom of the sea by a byssus cord. The animal spins fine mucus threads from a gland in the body, and the ends of the threads cling to the hard surface. The whole thread then hardens upon contact with the salt water, and the shell becomes firmly anchored for life. For their permanent home these oysters choose some solid surface at a depth of from 90 to 120 feet, and are harvested by divers who must cut them free one by one. The harvesting is government regulated. To prevent depletion of the beds, they must be allowed periods of ten to fifteen years for undisturbed growth.

Always with a multitude of enemies to battle, it is easy to see how only a few great pearl oysters could survive so long a life-span. And, with the strong appeal of their outstanding beauty, the demand habitually exceeded the supply. Quite naturally then, people began to think of the possibility of farming the great pearl oyster, just as they had successfully farmed the jewel pearl oyster. A British marine biologist, Dr. Cyril Crossland, began the first commercial mother-of-pearl farm in 1905, and devoted the rest of his life to making it a success against incredible odds.

At the time he began the operation, in which he had unlimited faith, Dr. Crossland was an employee of the Sudanese government. After an extensive survey, he determined that Dongonab Bay on the Red Sea was the ideal location for his experimental laboratory. He built his laboratory on the bay shore under near impossible conditions in order to, first, study the natural life history of the oyster, and then to devise a way to cultivate and produce tons of oysters where only pounds had grown before.

Dongonab Bay is in a dry desert area where several years often pass without a drop of rain. The nearest post with fresh water available was a five-day trackless ride by camel across the hot desert. For the three summer months he had to endure alternating days of suffocating sandstorms and steaming humidity.

After fifteen years of physical torture, patient research, and time-consuming experiment, he finally, by 1920, was producing three hundred tons of mother-of-pearl per year, where but seven tons had grown before. Then came World War II, Dr. Crossland's death, and the refusal of the Sudanese government to further finance the project, all traces of which were soon erased by violent sandstorms.

It was only through the persistent effort of his widow that the Sudanese government years later, in 1957, appealed successfully to the United Nations for help. A new laboratory was built on the site of the old on Dongonab Bay, and, using the knowledge gained by Dr. Crossland during his torturing years of experimenting, it is once again the only productive mother-of-pearl farm in the world, although many unsuccessful attempts have been made to establish others in various areas of the South Pacific.

The demand continues to greatly exceed the supply, and has prompted much chemical research to produce synthetic mother-of-pearl. Though many lovely artificial products have been made, demand for the beautiful natural nacre continues, for the lowly mollusk in its watery workshop has defied man's every attempt to duplicate, or even imitate, the luster of the genuine mother-of-pearl. The real nacre imparts a quality to jewelry, furniture, and other ornaments that can be gained in no other way.

Basically, the cultivation of the mother-of-pearl oyster involves the collection and protection of the very young during their free-swimming stage so that they can survive in greater numbers. Until their shells mature to some degree of thickness, they are highly vulnerable to fish, crabs, and many species of carnivorous mollusks. After they pass this vulnerable stage, they are placed in wire net beds where they grow to marketable size. The mortality rate is thus reduced to less than 10 percent, as compared with a more than 90 percent death rate under natural conditions.

The procedure is figured down to a science. Soon after spawning, floating rafts (spat collectors) are set in the sea. These collectors consist of many shelves of split-bamboo poles held certain distances apart in large wooden frames. The crevices between the bamboo offer ideal protection for the larvae until they become firmly attached. The rafts are suspended from floats not less than twenty-four feet above the sea floor to keep them out of reach of the multitude of sea-floor enemies.

After about six weeks, the spat can be seen by the naked eye. They soon grow to fingernail size and assume all characteristics of the adult. Another few weeks of growth and they will cover the spat collectors so densely there is no room for further growth. The collectors are then removed from the sea, dismantled, and the individual bamboo pieces, each with twenty to one hundred spat attached, are placed on a multilayered nursery tray, so spaced as to allow a year's growth without further crowding.

These nursery trays must be carefully fenced to prevent the entrance of predators. Apart from occasional inspection, the nursery trays need little attention. By the end of their first year of life, the mother-of-pearl shells are strong enough to withstand the attacks of most predators. They are then transferred to the final stage of cultivation—the growing trays—and on those trays growing space is allowed for at least three years.

The growing trays are then placed in areas very different from where the mother-of-pearl oysters are usually found. Because they must have a solid area upon which to attach themselves, they usually grow on pieces of old coral. In these natural areas they are surrounded by many enemies and must also compete for food. So the growing trays are moved to areas of mud or silt bottom away from competition. There the trays are supported about twenty inches above bottom on iron posts.

They are left on the growing trays for several years until they have reached proper marketing size. Some shells are permitted to grow larger than a dinner plate, while others are taken considerably smaller. They are then removed from the racks, opened, the animal matter removed, and the horny margins cut away. The remaining large nacreous bulk of mother-of-pearl is then ready for export to various markets all over the world.

Carving these lovely shells into valuable art objects is an important means of livelihood for the people of Jerusalem, which is one of the biggest importers. Some of these craftsmen show remarkable skill with their crude tools. Many work sitting flat on the ground without even a workbench. They just hold the shell between their knees and work with little hand drills, saws, and files. These skilled artists are paid at the rate of about forty cents per day. During my visit to Jerusalem, I added two beautiful carvings to my collection. They are so perfect that, had I not actually seen the work being done, I could hardly have believed it was achieved with such crude hand tools.

16. Uses of the Mollusk: From Ancient Times to the Present

From the beginning of history shells have served mankind in many interesting and unusual ways. The famous Golden Fleece of Greek mythology is actually not a fleece at all, but is, instead, the beautiful silken gold threads with which the pen shell (*Pinna nobilis*) attaches itself to rocks. They are so fine that a cloth-of-gold cape woven from them can be drawn through a finger ring. That gold-colored pen shell still thrives in the Mediterranean area, and its golden attachment threads are still being woven into objects of interest for the tourist.

The lowly mollusk provided ancient people not only with cloth of gold, which is perhaps the finest and most beautiful cloth of all, but furnished as well the gorgeous royal purple dye with which the robes of royalty were colored.

The dye was extracted from the murex species, and is believed to have been processed in Crete as early as 1600 B.C. At that time sharp-eyed observers noted that the milky white fluid secreted by the mantle of the murex shell changed, when exposed to the sun, through a series of yellow, blue, green, red, and purple hues.

It was the ancient Phoenicians, however, who took full advantage of this knowledge to create a great industry by manufacturing the dye known as Tyrian purple, although the old Tyrian dye was not purple as we conceive that color, but represented a wide range of color from dull red to violet. The ancient method of extracting the dye fluid was to crush the murex shell, a production method so costly that only royalty could afford to use the dye. So royal purple became the symbol of royalty, and has remained so to this day.

The secret process of making the dye from the mollusk juices was lost in ancient times and then rediscovered centuries later. The industry was eventually established in the New World in the West Indies and in several places along both coasts of South America. It is a thriving business in Mexico, where the modern method of production is to collect the fluid ejected by the excited animal and then replace the animal in the water to continue its growth and reproduction. So now, after nearly twenty-two centuries, the lowly blind and dumb murex mollusk is still providing beauty for the world to enjoy.

Mollusks provided not only cloth and dye for the ancients, but also gave them ink with which to write and do their artwork. So perfect and so without equal is the sepia ink—the so-called India ink—of the mollusk cuttlefish that modern twentieth-century artists are still using it with great satisfaction.

From the beginning of time superstitions have played an important part in people's lives. Since shells were of great practical use to people in ancient times, it is not surprising that they imputed supernatural powers to these beautiful and useful objects.

Many people believed that fertility of both crops and people hinged on the proper use of shells. In the South Pacific farmers planted an egg cowrie at the end of each row of crops, lest the seeds remain infertile. The Maori women of New Zealand found it essential to wear a headband of shells until their first child was born, if all was to go well. In the Congo a fertility girdle made of cowries was worn by girls to ensure fertility and easy delivery. In the South Pacific it was a necklace that had to be worn during pregnancy to guarantee an easy delivery. A left-handed whelk in a Fiji Islands home ensures that a family will be fruitful and multiply. Even now, in modern times, many Asian women hold a cowrie in each hand to ease the pain of childbirth.

Ancient peoples saw shells not only as a symbol of fertility, but as a symbol of eternal life. In Togo and Dahomey, cowries were placed in the graves of the dead to ensure their ability to buy food in the Other World, and to pay the ferryman for rowing them across the "Great River." In ancient China shells were put in the mouths of the dead to ensure a happy hereafter. Three were placed in the mouth of an ordinary person, five in the mouth of a great officer, and seven in the mouth of a feudal lord. The ancient Egyptians placed small cowries on the eyes of mummies to ensure sight through eternity.

Shells have played an equally vital role in the wars, religions, and various human contacts during those past centuries. The South Sea Island chieftains had no chance of winning a war unless they wore a cerebula carving from the large *Tridacna* clam to battle. It is first placed on an altar to give it supernatural power, and it is given all the credit for the victory. However, if the war is lost, both the cerebula and the chief are blamed and feared. The chief is beheaded and his head, together with the cerebula, is buried deeply in the ground to prevent further tragedy. For the people of Melanesia, if a battle is to be won, they must first place rows of the cowrie-shaped white ovula eggshells on the bows of their canoes.

Not only can shells help win a battle, but they help celebrate a victory. The Nagas and Dahomey peoples of Assam and Africa respectively wore them for prestige purposes, along with the hair from their decapitated enemies. On the Solomon Islands the "killer's badge" is worn to celebrate a killing. It is made from the ovum-ovum cowrie, and is topped off with a long protruding feather. A young Solomon Island warrior will proudly wear it on his forehead after he has distinguished himself in battle and won the honor by killing his first man.

Shells play an important part in the dress and in the ceremonies conducted by various tribal chiefs. In one tribe beautiful royal drinking cups are made by studding a big green turbo shell with gems and mounting it on silver. Some chiefs wear a beautifully made "kapkap" on the forehead, or temple. It is a flat shell disk overlaid by an elaborate tortoiseshell carving. The workmanship on the one in my collection is intricate beyond belief. Some chiefs wear breastplates of elegantly carved mother-of-pearl. For centuries, the golden cowrie was worn by the tribal chief of the Fiji Islands as a symbol of royalty. Until the rule was relaxed, no one else was permitted to own or wear a golden cowrie.

Certain chiefs have called their tribes to worship for hundreds of years with the triton trumpet shell. The sacred chank shell of India figures importantly in the life of the Hindus, and for centuries has been blown to summon the attention of the gods. In Christianity, the scallop shell was used as a Crusaders' symbol and has been worn for centuries on pilgrimages to the Holy Land. On certain islands in the South Pacific a "pakuisia" carved from a *Tridacna* clam plays an important part in appeasing the gods in a child-sacrifice ceremony. The carving must be suspended from the back of the neck of the child about to be sacrificed. In ceremonies of certain African cannibals, it was necessary to attach shells to iron crosses in the medicine bags they used.

Shells even found a place in games of chance. In Ceylon, Burma, and many other places they were, and are, used a little as we use dice, with win or loss determined by whether the shell apertures fall up or down. For an American, the four-leaf clover is a good-luck symbol; for the Japanese, a strip of dried abalone meat serves the same purpose. With every Japanese gift, a strip of dried meat wrapped in pretty paper is included to bring good luck.

For just ordinary people all over the world, shells serve many interesting and surprising functions. Women for centuries have adorned themselves with anklets, bracelets, nose rings, and so on made from shells. And they are worn with the same pride as the earrings, pins, and so on made of shells that Florida tourists bring back to Indiana each spring.

Farmers in the Channel Islands string abalones on poles, and when the wind blows they flash light and frighten away the predator birds. The South Sea Island natives use shells to cut their beards. Pet caged birds all over the world whet their bills on cuttlebone taken from the cuttlefish, an odd animal that is not a fish at all, but a mollusk related to the octopus and squid. And, with products made from that same cuttlebone, people brush their teeth, and smear ointments and dusting powder over their collective faces.

Oyster farmers circling the globe meet heavy losses from smothering algae growth. Now they have learned to control the problem by sowing great quantities of algae-eating periwinkle snails over the oyster beds. So when you pay for your next bowl of oyster soup, remember to say thank you to the little periwinkle for its assistance, for, without it, you would have paid even more for that soup.

The next chicken you eat, remember it was the oyster shells in its gizzard that helped the chicken reach your table. Tons of ground oyster shells are used in the poultry industry each year. From the time the baby chick is ready for its first meal until it reaches your dinner plate, the old oyster mollusk plays not one but a double vital role in its life. First, chickens have no teeth and their food reaches the gizzard whole; so the gizzard is provided with pulverized oyster shells to grind the food in preparation for digestion. Then, as friction wears away the hard shell, nature converts its hard calcium into bone so that the little chicken can travel to your plate on strong sturdy legs.

Perhaps without being aware of it you have traveled miles over roads built with crushed shells, or you have used lime made from them, or high quality porcelain. Probably, during that last trip to the art gallery, you were unaware of the part shells played in contributing to your viewing pleasure. Ground mother-of-pearl from shells adds a beautiful luster to certain paints that cannot be had in any other way.

Necessity is the mother of invention, so, in many countries less privileged than ours, shells have been crowded into service in many ways. They drill, scrape, carve, and peel; and, tied to a stick, they even hoe the garden. They serve as liquid containers, teakettles, oil lamps, musical instruments, and foghorns. In the dining room they are dishes, drinking cups, knives, forks, and spoons. They are combs, buckles, buttons, and stocking darners. They even find their way to the nursery as teething rings for infants, and they catch the fish for the whole family to eat.

Among the most valuable possessions of the ancient Hawaiian fishermen were their cowrie and mother-of-pearl fishing lures. Different styles were made for different kinds and different sizes of fish. They perfected one special cowrie lure capable of catching octopuses on the sea floor. They would choose a cowrie of exactly the right size, and then toast it over a fire to give it a reddish tint, which the octopuses found appealing.

Shells even protect a man in love from other overeager beauties, and they provide whistles for the children that come along later. To announce to the females of the tribe that he is unavailable, the South Sea groom-to-be wears on his head a special shell band made by his betrothed. The bands are made of tiny rice shells, and the amount of work on them is unbelievable. I was told there are ten thousand shells in the band now in my collection.

After the days of protection by the ten thousand shells comes the marriage—then the children—and then Papa gets busy making whistles. Gum chewing by American children and whistle blowing by certain South Pacific children are comparable in popularity. For over two hundred years, their mollusk egg-case whistles have been called musical chewing gum. Five types of snail egg cases are used. Larvae are removed and the tough case dried and stained many brilliant colors. The resulting five types are called sea whistles, snake bottles, sword whistles, war fan, and pigtail. So popular are they, that the snails are raised in bamboo pens to produce the cases commercially.

In a Tokyo cemetery a granite monument has been erected in memory of the seashell egg case. On its back are the names of forty-three whistle merchants who had it erected because they became conscience stricken over making their fortunes by killing snails.

Some few specific shell species are of such great value to the world that they deserve special mention. Artifacts dating back to 3000 B.C. made in the likeness of shells were found in Chile; and shells are still favorite models for modern artists. The sundial shell and the little *Latiaxis* temple shell hold a special place of honor for their great contribution to architectural design. The classic spiral staircase, designed by Leonardo da Vinci, is believed to have been patterned after the sundial shell; and it is widely supposed that the temples and pagodas of the Orientals take their shape from the little temple shell.

Another shell worthy of special tribute is the cameo shell. All over the world women wear and cherish beautiful cameo pins and rings, yet few people realize they are made from seashells. Any shell that grows with two or more distinct color layers can be carved, using one layer as a contrasting background for the figure carved in the other. Most such shells are either too thin for a full expression of design or they do not hold their color well when worn in sunlight. The one shell ideally suited for cameo work is the heavy three-layered helmet, which in years past was found abundantly in Italy.

With the material close at hand, Italy became the cameo-carving center of the world. A school was established in Rome for would-be carvers, who came from many parts of the world to learn from the old masters. For a number of generations the art was handed down from father to son. Then came the more rapid pace of twentieth-century living and, in their mad scramble for money, the younger generation no longer had the patience to learn this painstaking art, which is now dying out. The large lampshade-sized shells carved by the old masters are now so scarce that I worked ten years finding one for my collection, and they have become worth a fortune.

Another shell, the windowpane shell (*Placuna placenta*)—a wafer-thin shell about six inches in diameter—is becoming more and more widely used in the United States. For centuries, before anyone even knew what window glass was, it was used for windows in the East Indies. The shells are translucent and permit light to enter, although they are not transparent. They were the first ancient windows devised, and are still in popular use.

A large percentage of both lowly and elegant houses in the Philippine Islands still use shell windows. They are ideal to filter out the glare of the tropical sun, and make the use of curtains and blinds unnecessary. They are the most satisfactory material known to survive hurricanes, and they are ideal for bathroom and other areas where light, but no vision, is desired. In Manila alone, seven million windowpane oysters are used annually in house construction. Their use for construction work in the United States is now becoming popular. In a Bonwit Teller store in Pennsylvania there is an 88 by 15-foot wall spectacularly covered with these shells. They adapt beautifully to so many uses. In my collection is a table mat, which is a beautiful thing to see.

The shell is called the *kapis* by the Filipinos, and occurs in great abundance in bays. They are now being farmed, and the beds are contained by walls of bamboo. Growth is so rapid that only six months are needed from planting the young to harvest time.

The wafer-thin animal is made into omelets by the native people by adding milk, flour, and pork lard.

Perhaps the most popular and widely known use of shells we have not yet mentioned—shells as money. Of all the shells in the world, perhaps the little money cowries (*Cypraea moneta* and *Cypraea annulus*) are known by the most people. A number of shell species were used as money, but the cowries were most popular because no laborious manufacturing process was needed as with the other types. The cowrie was extensively employed in the South Pacific in the fourteenth century. In the seventeenth and eighteenth centuries it was the currency used by slave traders. Twenty thousand to one hundred thousand shells would buy an African slave.

When the cowrie money was first introduced into Africa, two shells would buy a wife. Later, with inflation, a bridegroom would have to fork over 2,500 shells. A cow was worth 2,500; a goat, 500; a fowl, 25; and a 62-pound ivory tusk brought 1,000.

As late as 1801 revenue from the British district was collected in cowries. A gentleman in the early nineteenth century paid for the erection of his house entirely in cowries. Sixteen million shells were needed to pay for a house worth $1,200. In 1873 the ship *Glendowra* was wrecked on the Cumberland coast of England. Its cargo of hundreds of sacks of money cowries was scattered for miles along the coast.

Another popular shell money was the wampum made from the quahog clamshell, and used from Maine to Florida by the Indians. The wampum beads were strung in six-foot lengths called fathoms, and they were woven into belts as tokens to bind treaties. The most famous treaty belt was the one given William Penn to record the land treaty made with the Pennsylvania Indians.

The Indians along the Pacific Coast of the United States used the long tusk (*Dentalium*) shells for their money. They were strung on threads and a six-foot length of the largest and best shells was worth $50.

In the South Pacific area many kinds of money were made from various kinds of shells. The most valuable was the heavy, thick doughnut-shaped disk carved from the huge five-hundred-pound clam (*Tridacna gigas*), used for the purchase of wives. If the girl was of sufficient beauty and rank, several disks were demanded. A young man, sufficiently eager, sometimes borrowed so many disks from his relatives that the rest of his life was spent laboriously carving enough replacements to repay his debt.

* * *

The physical shell itself has played an important role in man's history, but let us not neglect the equally significant, if not more so, place of the whole mollusk animal. So intricate and complicated is the structure of not only the shell, but the animal inside, that scientists, mechanics, and chemists have worked for years and are only just beginning to solve, and apply, a few of their hidden mysteries.

Some of the mechanical principles nature invented and built not only into her mollusks, but also into other sea creatures, have now been copied by man with great benefit.

The principle of the hydraulic suction pump was first perfected to operate the suction cups of a starfish's walking apparatus. The piston-vacuum principle that now solves many problems for man was designed by the Creator for squid. The smoke screen so useful in war is but a copy of the amazing built-in defense of the octopus and numerous other sea creatures. The octopus' ink sac expels a smoke screen containing a paralyzing agent that impairs the enemy's sense of sight and smell. The ball-in-socket joints so useful in many types of structures and machines was first invented and installed in the hinge of the *Spondylus* mollusk. The principle of jet propulsion now so common in man-made machines is a mere copy of the great Creator's plan of locomotion for the squid and octopus as discussed in the cephalopod chapter. Chiton mollusks and horseshoe crabs were both traveling around beneath the protection of armored tanks for many centuries before man copied the idea. And fish and birds were both gliding through the air all down through the years before man finally figured out the fundamental principle involved, and invented the glider. The file, corkscrew, and door hinge are all copies from mollusk shells. Even the principle of the gun trigger and hammer is copied from the fighting fin arrangement of the little triggerfish. The little frogfish had used a small built-in lure dangling above its mouth to go fishing for its dinner long before man copied the idea and began building artificial lures to do his fishing.

The mechanics have seemed to progress faster in exploring the marvels of nature than have the scientists and chemists. But the latter are now hard at work, and little by little they are making amazing discoveries, discoveries that lead one to believe that the cure for many of the diseases of mankind has for centuries been hidden within the living body tissues, glands, and secretions of ocean plants and animals, merely awaiting discovery.

After much research and experiment it is now hopefully believed that though a preventive for polio already exists, its actual cure is hidden in the abalone mollusk merely awaiting extraction, and that cures for diabetes and allergies have, all down through the centuries, been concealed in seaweeds merely awaiting the ingenuity of man to discover them. The lowly barnacle still holds the secret of the perfect dental adhesive, and now efforts are being made by the most brilliant of scientists to figure out nature's recipe for glue. Someday one of them will perhaps succeed and the whole world will benefit. Experiments now lead to hope that the cure for cancer is now hidden in land snails, and that relief for gout is to be found in ray fishes, while sea slugs and sea sponges are merely waiting for man to remove the cure for tuberculosis from them. The octopus is now believed to be a regular medicine factory for human ills. If its secrets can be successfully solved, it is believed to hold the perfect blood-clotting agent, as well as the cure for major heart disease and some kinds of mental illness. Also, as we discussed in Chapter 13, important insights into the structure of the human nervous system have been made by studying the nervous system of the octopus. In the same chapter it was pointed out that much has been learned about the human eye through study of the eye of the squid.

The lowly mussel, because of its ability to absorb poisonous materials from water, is now lending valuable help to scientific medical research. Since it is a filter feeder, it draws huge quantities of water through its gills to strain out minute particles of food suspended in the water. In the process it also strains out toxic poisons that build up in its body tissues. It has now been learned that a very powerful local anesthetic (100,000 times more powerful than procaine or cocaine) can be made from mussels that have been in toxic waters and eaten toxic organisms.

All filter-feeding mollusks have the ability to accumulate poison in this way. And through this ability, the clam is now lending aid to pollution experts in Illinois, Indiana, Wisconsin, Michigan, and Minnesota. A way had to be found to filter DDT from the waterways and measure it in accurate proportion to its amount in the water. For a time scientists were stumped because DDT is practically insoluble in water and could not be measured. Then they thought of the uncanny and superior filtering ability of the clam, and the lowly clam is now going to do what no man-made gadget could do. Clams are to be shipped into the area, placed in the water, and each week scientists will cut them open and measure the amount of DDT they have filtered from the water and absorbed into their body tissues. Thus, the danger point can hopefully be recognized in time to prevent a tragic poisoning of humanity. What a huge and fascinating task scientists do face as they struggle to unravel the mysteries wrought by the great and all-wise Creator.

PRACTICAL SHELLING

17. Practical Shelling

Seashells, which are among God's most beautiful creations, inhabit every ocean throughout the world from the Arctic to the equator in both hemispheres, and from high-tide line to the very depths of the ocean. Inconsiderately, however, nature has arranged it so that the most beautiful and interesting species do not lie around in plain view. But with the pointers in this chapter, a beginning sheller will have the know-how to set out upon this most satisfying and rewarding activity.

Where and How to Hunt

As you begin your search, do not let the beautiful cleaned shells you have seen on display mislead you. Nature has many disguises. In their natural state shells are often so skillfully camouflaged that you can only recognize them by the shape of the queer object you see on the beach (Plate 36 a–d). Many people leave a beach convinced it is barren merely because no pretty cleaned shells were lying about in plain sight as they had expected to find them. Or they may leave disappointed because all they found were dead, faded, broken, beach-worn shells. Many people fail to even find the beach because, unknowingly, they arrive at high tide when the beach and all its shell life is covered with water. When the tide goes out, all the shell creatures on the then-exposed beach do an amazingly clever job of hiding from their enemies and from the direct rays of the sun. So cleverly do they hide that, to the inexperienced collector, the beach often appears quite barren although it is teeming with shell life.

The best time to hunt is during the spring months' minus—that is, extremely low—tides. Florida shelling reaches its peak from early February through late May.

To find the deepwater species you must, of course, dive or dredge, but hundreds of beautiful species can be found by searching the beach and by wading out at low tide. An early morning search of the ridge of debris deposited by the high night tide can be very rewarding, especially after storms. But the real beach search must be conducted all the way from that high-tide ridge down to the water's edge at low tide—and on beyond as far as you can wade. (Assuming you are a nonswimmer, you must be content to stop at that danger point.)

The mud flats must be thoroughly searched for telltale signs where shells have dug in for safety. For example, tracks on the sand must be followed to trail's end where the shell can often be flipped out of the sand where it has gone into hiding. Tide pools must be carefully gone over, preferably looking through a glass-bottom bucket if the shells are very deep. Water spurts must be dug into with your shovel.

All rocks, driftwood, and so on that offer hiding places underneath must be rolled over. If they are too heavy to move, the cracks and crevices must be probed into. Many mollusks can be found hidden among, and clinging to, seaweeds. Often a good live shell will take refuge in a larger, dead empty shell.

Another method that sometimes reaps rich rewards is to lay out rotten meat, fruit, and so on to attract hermit crabs. They will, of course, bring with them to your banquet all the empty shells they are living in. Ofttimes one of their adopted shell houses is worthy of a display cabinet. The famous Edwards glory-of-the-sea (found on the beach at Rabaul, New Britain, and now in the Australia Museum at Sydney) was believed to have been carried to the area by a hermit crab attracted by the bait Mr. Edwards had put out.

The search for miniature species—that is, very small species—requires a different technique that must be considered separately. With magnifying glass and tweezers at hand, many special places must be searched. The ridge of sand and gravel around rocks is often very rich in miniature species, as is also the debris from the slots between rocks. Scoop up dredgings from beneath the water and carefully search with the magnifying glass, and look carefully into tide pools with it.

Stir sand from the high-tide ridge of debris into a bucket of water. Spread the top skimming on a dark cloth, and pick out the tiny shells with tweezers and magnifying glass. A fine mesh screen is also helpful.

Search anemones for tiny shell parasites. Shake residue from kelp holdfasts (or roots) that have been swept onto the beach and search them with a magnifying glass. Miniatures can also be found on eel grass and in sponges. Look for tiny riders on abalones and on other large shells. Search starfish stomach contents and also the outer surface of starfish, sea urchins, and sand dollars.

Every sheller has his own favorite shell-carrying utensil. After considerable experimenting, I like a large, lightweight, flat, wide-bottomed aluminum kettle best. Also take a small rake, a garden trowel, and a sharp thin-bladed knife for limpets.

If the water is ruffled by wind, it is impossible to see bottom, and for a clear view a glass-bottom bucket is essential for tide pools and wading. It is helpful to wet the inside of the glass occasionally, and paint the inside a dull black to reduce the reflection of the sun on the glass. The important thing to remember is not to go so burdened with such elaborate equipment that it becomes impossible to carry the treasures you find. To be ready for those extra rewarding days, I also carry a large cloth or plastic bag. It can be tucked in belt or sweater pocket and carried with no effort, and it is a lifesaver at times when a way to carry an overflow is needed.

In general, remember to restore rocks, logs, and so on to original positions after looking underneath. Otherwise the eggs and very young will be destroyed, as they cannot survive direct sunlight.

Leave all damaged unusable shells in place so they can reproduce. Take only what you have a real use for and leave the rest for further growth and reproduction. Never collect merely for the fun of the catch and then throw them away. God put them there for a purpose—not to be merely destroyed and thrown away.

Remember to stay in good cheer, although you are unsuccessful. There are seasons for shells just as for birds and fish. Beaches may be filled, then months elapse before they are filled again. Perhaps the same thing will happen the next year—or perhaps not for several years.

Unfortunately, increasing civilization and development projects are making a serious inroad on the shell population. Many species are already extinct. Great scopes of beach areas of former excellent shelling are already entirely destroyed by modern development projects. It is indeed easy to understand the apprehension of scientists that shelling for our not too distant future generations may be only a past activity to read about.

Shelling in Safety and Comfort

With all the satisfaction of shelling, it is important to remember that there are some discomforts, and even hazards, involved. So a word of counsel at this point may not be amiss.

Basically, wear comfortable loose-fitting clothing. Let the hot sun, cold wind, or prevailing insects determine the amount of bare skin to expose. Modern quick-drying materials have a great advantage, and dull colors have the advantage of being less attractive to certain insects. Slacks or shorts are a favorite garb of both men and women. But for women who may be spending a full day or days on the beach, a comfortable dress of extremely quick-drying material is the ideal solution.

In choosing shoes keep four things in mind. *First:* Nonslip thick soles. *Second:* They must go on and off easily when wet. *Third:* They must have no openwork trim to admit fine sand. *Fourth:* They must be of a material that breathes. Your feet will soon be sore if encased day after day in water-filled rubber shoes.

Also, for feet (as well as for hands) carry a supply of Band-aid patches at all times. Apply at the very first sign of blisters; do not wait for them to actually develop. It is very difficult to continue shelling and try to heal an open blister under water.

Animals can be a problem from insects on up. Unfortunately, the richest shelling occurs during the hours that sand flies and sand fleas, not to mention mosquitoes, are out in force. These little creatures can cause excruciating torture, so a handy insect repellent should be applied often and liberally.

Barnacles, which encrust many of the rocks that may have to be lifted to discover shells underneath, deserve proper respect; heavy gloves, which are cumbersome, or, preferably, taped hands will give valuable protection. The tape or gloves will also protect against the heavy rocks themselves which can seriously abrade your hands as they are lifted. Coral causes a similar problem in tropical areas. Dead coral, especially, can easily break under pressure, peeling an ankle or leg in the process.

Beware, too, of the seemingly innocent blob of jellyfish, which live in all seas. Its sting is always painful, and sometimes deadly (depending on the species). Even jellyfish that have been washed ashore looking quite dead and harmless may still have the power to deal a devastating sting. See page 36 for how to proceed if you should be unfortunate enough to encounter one of these creatures.

The sea urchin, quite numerous in many areas, is another villain for the unwary sheller. Its sharp spines produce a poisonous sting that is both very painful and very long-lasting. The most successful pain remedy that I know of is to apply meat tenderizer.

One last dangerous creature, which inhabits exotic tropical areas such as Australia's Great Barrier Reef, is the stonefish. This ugly and venomous creature is so nearly perfectly camouflaged that it is next to impossible to see. Just keep your eyes peeled and fingers crossed.

Another set of safety problems while shelling concerns the constantly changing sea itself. Low tide may find you far out on a tempting sand or mud flat. But keep your eyes open for a rising tide, for you may be on a rise or sandbar. It is easy to become so absorbed that, without your noticing, the tide fills up the long trough between you and the shore. At best, you will have to swim back to shore or be marooned on the bar until the tide turns; at worst, where the difference between low and high tides is six feet or more, you will be stranded where the water will soon be over your head.

Tide traps can be a problem in areas where an ocean and mountains meet on a coastline. A mountain base does not protrude an equal distance into the ocean all along its length; if sections of the base in front of and behind you both project farther into the ocean than the spot where you are shelling, watch out. The water will get very deep very quickly in this small bay, and can cut off your retreat from either direction, once again stranding you where the water will soon be over your head.

Keep an eye on the waves, too. It is unlikely that you will be shelling out in the water on a windy day with very high waves, but even on days of moderate wave strength, the occasional sneaky high wave will come along and, if you are not careful, knock you out of your depth.

Here are a few more pointers to start you on your way:

On an all-day hunt away from the car, that inner man must be remembered. To keep him happy, I start out with a sack of bananas, two or three cans of fruit juice, and a can opener. The bananas are nourishing, moist, and do not aggravate thirst. And the emptied fruit-juice cans supply the need for separate containers (in the big kettle) in which to carry fragile or small shells.

On a windy day, blowing sand can sometimes cause painful injury (it's no fun to have shins that look and feel like they have been scraped with sandpaper). It is a good idea to tape the front of your legs before starting out.

Remember, before leaving home, to conceal safe pockets in your shelling clothes in which to carry a car key. Your purse or wallet must be locked out of sight in the car trunk, for you cannot carry it on the beach. Neither can you carry a car key in your hand, but you must take one with you to reenter your locked car.

And finally, be sure to note a prominent landmark at your point of departure as you begin the day's excursion; otherwise it is easy to lose your bearings while walking back at dusk.

And now you are ready for a beach excursion.* Happy shelling!

* For information on how to plan a serious shelling trip covering many miles of long-distance travel, see *Pursuit of the Wild Seashell*, the forthcoming companion book that tells of my travels and adventures.

Killing, Cleaning, and Storing Shells

Killing

The most unusual method of removing the animals from shells that has been called to my attention is singing to them. I did not see it demonstrated while there, but I was informed that the natives in certain areas of the South Pacific hold a living shell close to their lips and softly sing and whistle to it. Result: The charmed animal obligingly creeps from the safety of its home and permits itself to be captured by a quick hand.

Since the general success of this method would be considered somewhat dubious, we will proceed below with what may prove to be more practical procedures.

Bivalves

If placed in fresh water or in carbonated water, bivalves will die and open so the adductor muscle can be cut for cleaning. Some open quickly. Others require several days, and the water should be changed daily.

Bivalves will also open nicely if placed in a freezer. If small, such as coquinas, they can be left standing twenty-four hours in strong household ammonia water. Rinse several times and the small bodies will float away, leaving the hinges unbroken and the color undamaged.

Extremely small bivalves are cured whole by many collectors by letting them stand in rubbing alcohol or buffered formalin for a day or so. They can then be dried and placed in the cabinet with no odor developing. Although it is quick and easy, I very much dislike this method because of the powdery dust that sifts from the shells years later.

Bivalves are also quite easily steamed open in the top of a double boiler.

Many people boil bivalves, but I feel they are nicer when killed with fresh water—sunray shells especially should not be boiled, as they have a tendency to turn pink in hot water.

If the species has a byssus, it should be saved to be treated later with oil and preserved with the shell.

Univalves

Collectors living in hot areas often bury univalves in sand near an ant hill and the meat is soon eaten away. However, when using this method, great care must be used to have each shell standing on end so that all liquids will drain downward, away from the shell. Otherwise, the acids of decomposition will be in contact with the shell and the surface will become etched. If a little sugar is sprinkled on the animal or in the aperture, the ants will be attracted more quickly. All shiny shells such as cowries and *Oliva* should be protected by a coating of Vaseline before being buried near an ant hill.

Very delicate shells can be damaged by sand. Although the developing odor is most objectionable, some collectors in warm areas prefer to use the following method: Heavily line a box with corrugated paper, place shells in box on their backs with apertures facing upward, then set the box out of smelling range, if possible. Within a week or so the animals will decay in the warm temperature and can be flushed out with a garden hose.

For collectors on the move, boiling is one of the most universally accepted methods. Care must be taken to start with cold water and bring to a boil slowly, as a sudden change of temperature is harmful. Line the bottom of the pan with cloth to prevent direct contact with the heat source. If glossy shells, such as *Strombus*, *Cypraea*, or *Oliva*, must be boiled, it is helpful to pour some mineral oil into the boiling water, making sure the shells are entirely submerged. Then, immediately after the animal has been removed, rub a little mineral oil or Vaseline on the glazed part of the shell. The length of boiling time varies with the size and type of the shells. For some, just bringing to a boil will be sufficient. Others must be boiled two or three minutes, and often the large sizes must be boiled half an hour. For best results, let the shells remain in the water as it cools, then remove the soft parts as soon as they have cooled enough to handle. Use a small pin, bent wire, crochet hook, large darning needle, or an ice pick (depending upon the size of the shell), and pry out the animal with a twisting motion.

If they refuse to come out, try boiling a little longer. Check each animal carefully as it uncoils from the shell. Often the tip breaks away and remains in the shell. Sometimes it can be loosened by a quick jolt. If not, keep the shell separate for further treatment or you will rue the day. (See "Removing Odor.") Hose out the inside well, and remove and save all opercula, or doors, to replace later after the shell has been cleaned.

Very small univalves can be sealed in plastic bags or glass jars filled with water, and the decomposed meat will wash away nicely in a few days. However, the water must be carefully changed daily to prevent discoloration of the shells. Follow the final rinse with a Clorox rinse to remove clinging odors.

Very small univalves can also be put in fresh water overnight and then in a 70 percent ethyl alcohol solution or a 10 percent formaldehyde solution for two or three days. These preparations can be bought in any drugstore. Although few people undertake the rather tedious job, it is possible to remove the soft animal part after the alcohol treatment. But the formalin—that is, the formaldehyde solution—hardens the meat permanently until it is forever impossible to remove it from the shell. By this method of quick and easy cure, the opercula will stay nicely in place and all will be fine—for a while. But, after a while, the cured meat will attract insects and sift out as an ugly brown powder that makes the display area look dusty and dirty. Personally, I so strongly object to this method that the extra time and effort needed to remove the meat seems well spent.

For those who prefer to preserve the animal in the shell, it is important to let the animal die before it is placed in the solution so the operculum will remain nicely in view. Living animals withdraw deeply within their shells when put into a preservative, often with the operculum quite hidden from view.

If the shells are to be stored in the solution for any length of time, do not use formaldehyde, as it has a tendency to weaken and decompose some shells. They will not be harmed by prolonged contact with alcohol.

Large univalves can also be treated with alcohol to kill the animal for removal. Many collectors prefer this method to boiling. It has two distinct advantages. It permits the busy collector to postpone the time-consuming animal-removal task to a later, more convenient day. And for the needs of the scientist, the animal, if necessary, can be dried after the alcohol treatment and much later softened with a detergent solution for laboratory study.

To kill univalves chemically, put in fresh water, to which a little chloride or sulfate of magnesia has been added. When the animal expands and no longer draws back in when touched, put in either strong alcohol or an 8 percent formaldehyde solution to kill. Then remove the animal at once.

To kill by refrigeration, place the live shells in a deep freeze and leave for at least seventy-two hours. Often a much longer time is needed. The animal shrinks and the attachment muscle separates from the shell. Remove from freezer, allow cool (never hot) tap water to flow over shells for ten or twenty minutes, and the soft parts can be pulled, cut, and hosed out. Tropical shells can be killed just in the refrigerator, but the deep-freeze temperature is needed for cold-water species.

Shells containing hermit crabs should not be boiled. It is all but impossible to remove the killed crab. Put those shells in fresh water with a little Clorox. The crabs will leave the shells and die outside.

Cowries should never be dropped in boiling water, as it often causes their enamel to check. If boiled at all, drop in very cool water. Many people do boil them, but I feel they are all damaged a little by that method. I feel it is far better to put them in a deep freeze for a week, if possible. If that is not possible, inject hydrogen peroxide into the animal with a hypodermic needle and the animal will fall out. Or, if all conditions are favorable, stand them on end in sand and leave them until the decayed matter drains away, as the ants clean them up.

Cones, and other shells with small apertures, should be soaked two or three hours in a solution of Epsom salts (one teaspoon to one pint of seawater). Otherwise the animal retreats completely out of sight.

Murex are better if the barnacles and so on are scraped away before boiling. The heat makes the spines more brittle and seems to harden the incrustations to make a later cleaning more difficult.

The horse conch periostracum seems to survive boiling water if it is first washed well in salt water.

Carrier shells present a special problem since the body retreats out of reach when boiled. Let them lay exposed to air a day or so until they die. Then boil and remove soft parts.

Strombus shells of the highly colored species often fade if boiled. They can be most successfully killed by the deep-freeze method. The larger species require as much as two weeks in the freezer.

Oliva shells have a tendency to lose luster if boiled and are much better killed with alcohol or by freezing.

Terebra shells need also to be refrigerated or else treated with Epsom salts. Merely boiling them causes the animal to retreat out of reach.

It is best to keep the shells alive until cleaned or frozen. If kept wrapped well in wet rags or seaweed, they will live for two or three days. If a longer delay is necessary, salt them heavily and they will be less offensive.

Cleaning

Removing the lime, worm tubes, barnacles, sponge growth, algae, and so on from the outside of shells forever remains a matter of experimenting. No one can tell another person how to cook, iron, paint a house, or do anything else requiring skill gained through experience. They can only offer suggestions. It is my hope that the hints listed below will prove helpful.

Clorox is very helpful for many things. For colored shells it must be diluted fifty-fifty or more and the shell must not be soaked too long or it will damage the color. For white shells it is an excellent bleach used full strength. Most white shells can be left in overnight. But better keep a close watch and study reactions before you risk an overnight operation.

Pineolum seems to be a favorite cleaner on the Hawaiian Islands. Its oil base prevents harm to color and gloss, and the shells can soak in it longer. So far I have been unable to locate it in the continental United States.

Caustic soda (or lye)—one pound to one gallon of water—as an overnight soaking solution will sometimes remove calcareous deposits and barnacles that resist other methods. It also removes the periostracum, or epidermis, which is good or bad, depending upon your objective. The lye is, of course, injurious to hands or to metal containers and must be used with proper precautions. But it will not damage the color of the shells, as can happen with the use of commercial bleaches. Extremely encrusted shells can be left in the solution for as long as forty-eight hours without damage. Samples should be treated to determine the proper timing.

Chloride of lime is effective in removing the rough outer coat. First wash surface and with a strong knife scrape away as many incrustations as possible. Moisten lime with water to form a smooth paste. Spread paste over shell surface with a soft brush and let stand several hours. Examine periodically to make sure the paste is not eating too deeply into the surface. After the covering is eaten away, wash well in warm soapy water, and the surface should then be exposed in a glossy condition.

Snow White used full strength for a twenty-four-hour soak will not hurt color or gloss and will sometimes remove lime and other matter that resists other methods.

Muriatic acid is very helpful in many cases but requires much experience or experimenting as to percent of dilution and length of application. It can destroy the delicate sculpturing, or even eat the shell away until it entirely disappears, if used too strong or too long. One recommended method is to dip the shell one second in the full-strength acid, then plunge immediately into ice water with the ice cubes still floating. Repeat if necessary or finish spots with a pastry brush. A second method is to place shells in a mesh bag, shake in a solution of three ounces of the acid to one gallon of boiling water, then wash well in warm soapy water. A third method is to use one teaspoon acid to one pint of water. Brush on spots or dip quickly. Then wash in a strong baking soda solution. Repeat as often as the change is for the better.

A colorful aperture can be protected from acid damage by filling it with water and immersing the shell only to the level of the aperture.

Warning: Avoid use of all metals in working with the acid, as it will ruin metal containers or utensils.

For the very dirty shells (especially mussels), soak for a week in two parts Energine and one part light mineral oil. It does not seem to damage the shells and is sometimes effective when all else fails.

Elbow grease seems to be the only solution a large percentage of the time—just hours and hours of scraping and scraping, of cleaning the shells inch by inch with good old elbow grease, seems to be the *only* way to salvage many a beautiful shell. If one has but one of the species, and if a last suspicion remains that a thing of rare beauty is concealed beneath, no effort is too great, or too time-consuming, to complete the job.

Removing Odor

In spite of one's most careful effort in removing the animals, often a portion will break away and remain lodged in the tip of the shell. The smell is one of the hazards that must be endured by all collectors.

Some prevent the strong odor from

forming by soaking the shell in alcohol or formaldehyde, but eventually the preserved chunk of meat will attract insects and sift down in a dirty brown powder to spoil the appearance of a showcase.

A far better method that often works is to stand the shell on end, fill the aperture with full-strength shampoo or detergent, and let stand a few days. That will often loosen the offending chunk so it better can be shaken out. All other efforts failing, the best final results of my experimenting have been obtained by filling the apertures with melted paraffin. Let it harden and then pour in more to seal the edges. It does quite a perfect job, confining both odor and dust. Pouring in shellac is sometimes also satisfactory.

Polishing or Oiling

Polishing or oiling is a very sensitive subject among collectors. Many collectors prefer that the shells remain in a completely natural state, although color or finish may be dull. Many others feel that, especially with the duller species, an artificial polish, or gloss, is an improvement. Consequently the sale of many a shell is lost when collectors belonging to the two schools of thought attempt to deal with each other.

Most live-taken shells have such a beautiful natural gloss it is hard to see how anyone could feel tempted to undertake an artificial improvement. However, some, although live-taken, have a disappointing dullness of color due to the loss of some of the natural oil in the shell. The addition of oil in those instances is approved by some collectors, though not by all, although the oil replacement does brighten the color somewhat and seems to somewhat lengthen the life of the shell.

The thing most violently disapproved by the highest percentage of people is the addition of lacquers, or varnishes, in an attempt to simulate the high gloss of nature. Once the high gloss built by nature has been lost, no power on earth can replace it and make the shell acceptable in a selective collection.

Sometimes shells are polished on emery wheels, not to restore a natural gloss as by using varnish, but to grind away an outer layer to reveal what is underneath. Having single specimens of certain species treated by that process is essential for an educational exhibit such as mine. Some species take that type of polish exceedingly well, and specimens of great beauty are obtained by revealing that inner construction.

Each collector who uses an oil replacement to restore a color loss has his own favorite recipe. Some of the most commonly used are: One part mineral oil to two parts Energine; pure 3-in-1 oil; one part mineral oil to three parts mineral spirits; pure mineral oil; Windex; one part mineral oil to three parts chloroform; full-strength lemon oil.

My favorite recipe is one part baby oil to ten parts lighter fluid, using just the oil on the color blue, since the fluid will make the blue color run, especially blue coquinas.

A treatment of furniture wax imparts a nice finish to some shells. It leaves an unnatural look, but some collectors coat their shells with silicone plastic to stop light-ray damage.

Some hair-dressing preparations rubbed into some shells leave a nice finish.

Some collectors soak their cleaned shells in water to which mineral oil has been added, believing it makes them a little less brittle.

Bivalve hinges: A drop of glycerin on a bivalve hinge will prevent the hinge from breaking. The two valves should then be tied together unless the open shell is preferred.

Repairing

Sometimes a valuable and otherwise flawless shell has a small hole, especially the lion's-paw pecten. The hole can sometimes be mended with a small piece from a perfectly matching crayon.

Unsightly white spots at the broken tips of starfish projections just seem to vanish if touched with the proper shade of art paint.

Transparent airplane cement often does an excellent job of repairing a shell that is broken in handling. And slightly chipped lips can sometimes be repaired with fine sandpaper.

Storing

Shells should be stored away from dust and direct sunlight. Total darkness is best for some species. They will in time lose color from the fading action of the sun. And a coating of dust causes a rapid and fatal deterioration of the shell.

Mildew preventatives and mothballs should be stored with starfish, crabs, and so on.

Wads of cotton dipped in formalin will help prevent fungus growth if put in the boxes. Plant and animal oil promote fungus growth, so mineral oil is recommended. It is the collectors in humid areas that are reluctant to use oil of any kind since its addition tends to encourage the growth of fungus. Fungus is a plague in tropical climates, and the only method of perfect control is to never let it get started. Surface fungoid growth can be wiped away with a cloth, but once it has eaten into the shell enamel, no amount of washing or rubbing will remove it. It is high humidity rather than high temperature that is conducive to fungus growth. During the tropical wet season, specimens should be wiped over regularly at least every three weeks unless chemical preventatives are used in the storage cabinets. If the drawers are airtight, the addition of moisture-absorbing silica gel is helpful. It can be reused again and again if dried out in an oven each time it becomes saturated. If the cabinets are not airtight, it is recommended that tiny holes be bored in the bottom of each drawer and light bulbs placed beneath the cabinets during the periods of high humidity.

To prevent cracking, coat shell with equal parts of glycerin, alcohol, and water. This is especially good for some species of land shells. A drop of pure glycerin on a bivalve hinge will keep it less brittle.

Bonding Agents

Airplane cement holds shells the most permanently and solidly of any bonding agent within my experience and I have tried several. Use only where you want the attachment to be permanent. It can be removed from the fingers with nail-polish remover or with pure acetone.

Du Pont has a special plastic cement with which shells can be attached to a modern plastic surface.

Tiny shells can be attached to glass with gum tragacanth, a dry powder. Mix with a few drops of water to form a jelly consistency. Water will resoften it so the shell can be removed.

Preserving Small Animals

Specific rules for special things work best as explained later. But as a general procedure, for miscellaneous small life such as anemones, jellyfish, worms, small fish, sea cucumbers, and so on, the following method works very well. Put the animal in a fully concentrated solution of Epsom salts to get it in a relaxed state of expansion. Then place it for several days in a formalin solution made by mixing one part of commercial formaldehyde with nine parts water. Store the jellyfish and like soft-bodied creatures in jars of fresh solution. Solid bodies like starfish should be removed and dried thoroughly. Colors fade and are often replaced by skillfully applying artist's paints. Many people spray with liquid plastic. That ensures an excellent state of preservation but creates an artificial appearance.

Crabs are more liable to retain their color and not shed their legs if the formalin is injected directly into the soft parts of their body. Then a few hours later drop them in a glycerin solution made of equal parts glycerin, alcohol, and water. Let them set a few days, depending on size—an extra three or four days for the larger ones. Remove and keep in a dark place until thoroughly dry.

Sand dollars will die and dry if placed in a cool safe place. Their spines can then be rubbed off in warm water if preferred. Many people prefer them in the completely natural state. If a white color is preferred, soaking in Clorox will help—but not too long or they will readily crumble. Carefully break apart at least one specimen to retrieve the small "doves" inside.

Sea urchins are almost impossible to cure with the spines erect as in life. But the best method I know is to lay them on the beach instantly (mouth down), and quickly cover them with a mound of damp sand to hold the spines as erect as possible. Cover with a cloth and flip over. Draw cloth poke together and tie, leaving opening a little wider than the mouth. Remove all animal matter from the hull through that opening. Buy some 20 percent formalin solution at your drugstore and fill the shell with the solution, letting it stand for twenty-four hours. Hose off and brush spines with a soft long brush to bring out gloss as they dry in the shade. Many people then spray with varnish or plastic to hold the spines in place for a longer time. It perhaps does serve that purpose, but it leaves them looking very artificial. So far as I know, no way has ever been found to preserve the full brilliance of their original beautiful colors.

Chitons—First method:

Remove animal and skin covering inside the plate immediately. Bind tightly to a flat narrow board, thus forcing it to dry flat. If it has already curled in a ball in the collecting bucket, put it in a container of seawater and it will straighten out for cleaning.

Chitons—Second method:

Tie flat on a flat narrow board and drop it in seawater to which Epsom salts has been added. When it is dead and the body swollen flat, untie, remove the body leaving the girdle intact, retie, put in a fifty-fifty solution of glycerin or 10 percent formalin, and let stand twenty-four hours. That method leaves it slightly pliable.

Chitons—Third method:

Tie flat on a narrow board and drop in alcohol or formalin solution for twenty-four hours. Then remove soft parts and inner lining with a sharp knife. Chip all crustations from the outside. Then soak for several days in a fifty-fifty solution of alcohol and glycerin. Bind again on a flat stick to dry. If the girdle shows signs of curling in a day or so, coat it with mineral oil.

Chitons—Fourth method:

To preserve just the chiton valves, rot or boil them until the girdle can be removed. Before removing it, carefully number each valve or else carefully place them in the right order, as they are removed. Then glue them back together.

No matter which of the four methods is used, before the final tying to dry, brush the top of the chiton well to remove all dirt and crustations.

Starfish—First method:

Relax the starfish in a warm weak solution of alcohol and water. Turn them upside down and, using a medicine dropper, drop formaldehyde along the arms. Let them rest for a few hours, then drop them in a solution of equal parts glycerin, alcohol, and water. Let stand several days, depending on size. The larger ones take longer. Drain on absorbent paper on a flat surface. Dry in total darkness for best color results.

Starfish—Second method:

Wait for the starfish to flatten out in the bottom of a pan of seawater. Turn on the hot-water tap and slowly change the seawater to warm fresh water. The water need be no hotter than you can bear your hand in. It kills and fully relaxes the starfish at once. Pour off the water and add 50 percent formalin to fully cover it. Arrange the arms as desired and let it soak for several days. Slide out on a flat board and hose all dirt away. Put on dry board to finish curing. Tie the arms down if necessary. Dry in *total* darkness to prevent colors from fading. Layers of cloth can be tied around it to create darkness. Hang in a shed to dry. It may take several months to dry thoroughly but it should retain its original color quite well, although I have yet to see the first cured starfish that retained the brilliance of the natural living color. Cured starfish can be, and many are, quite skillfully painted to an approximate original color.

Turtle shells are often a great disappointment. If, after the animal has been removed, the shell is just laid out in the sun to dry, it has a tendency to warp, crumble, or come apart. This can be prevented if the freshly cleaned-out shell is packed in borax (just the usual laundry borax) and dried more slowly in the shade. The borax draws out the oil and leaves the shell more stable. When it is thoroughly dry, carefully brush away all the borax.

Many turtle shells take on an attractive gloss if buffed. It leaves an artificial appearance, to which many people object, but the shells also take on a brighter gloss if shellacked or sprayed with a Krylon preparation now available in art stores.

Measuring Shells

Many shells are far too small to measure in inches, and often the larger shells must be measured with greater accuracy than can be achieved using fractions of an inch. Thus, it becomes necessary to use the millimeter as a unit of measure.

Fractions are involved in reducing the unfamiliar millimeter so we can think in terms of the more familiar inch. For a simplified quick comparison, I have worked out the approximate table below. Although it falls short of complete mathematical accuracy, I have personally found it very helpful.

3 mm = ⅛ inch	25 mm = 1 inch
6 mm = ¼ inch	50 mm = 2 inches
12 mm = ½ inch	75 mm = 3 inches
	100 mm = 4 inches

Selected Bibliography

The larger part of my research on this book I carried out on my own, traveling around the world with pen in hand, but the final product could never have been accomplished without the help I gleaned from the work of others. Though a complete listing of every valuable source would be impossible, I offer here a sampling of the books and periodicals I consulted, with my thanks for the insights and information they offered me.

Abbott, R. Tucker. *American Seashells.* New York: D. Van Nostrand Co., 1955.

———. *How to Know the American Marine Shells.* New York: New American Library (Signet), 1970.

Abercrombie, M.; Hickman, C. J.; and Johnson, M. L. A *Dictionary of Biology.* Baltimore: Penguin, 1964.

Allan, Joyce, F.R.Z.S. *Australian Shells.* Boston: Charles T. Branford Company.

Borror, Donald J., and Delong, Dwight M. *An Introduction to the Study of Insects.* 3rd ed. New York: Holt, Rinehart, and Winston, 1970.

Burghart, Glen and Laura. *West Coast Chitons.* San Francisco: San Francisco Aquarium Society.

Carson, Rachel. *Edge of the Sea.* Boston: Houghton Mifflin, 1955.

———. *The Sea Around Us.* Rev. ed. New York: Oxford University Press, 1961.

Cernohorsky, Walter O. *Marine Shells of the Pacific.* San Francisco: Tri-Ocean, 1967.

Engel, Leonard, and the editors of *Life. The Sea.* New York: Time-Life, 1967.

Gilbert, Keith, and McNeill, Frank. *The Great Barrier Reef and Adjacent Islands.* Sydney, Australia: A. K. Murray Pty. Ltd.

Graham, E. C. *The Basic Dictionary of Science.* New York: Macmillan, 1966.

Gray, Asa. *Manual of Botany.* 8th ed. New York: Van Nostrand Reinhold, 1950.

Gray, Peter. *Dictionary of the Biological Sciences*. New York: Van Nostrand Reinhold, 1968.

Heron-Allen, Edward, F.R.S. *Barnacles in Nature and in Myth*. New York: Oxford University Press.

Hickman, Cleveland P. *Integrated Principles of Zoology*. St. Louis: C. V. Mosby Co., 1966.

Hirase, Shintaro. *Shell Handbook in Natural Color from the Japanese Islands and Adjacent Territory*. Tokyo: Bunkyokaku.

Johnstone, Kathleen Y. *Collecting Seashells*. New York: Grosset & Dunlap, 1970.

———. *Sea Treasure*. Boston: Houghton Mifflin, 1957.

Keen, A. Myra, and McLean, James H. *Sea Shells of Tropical West America*. Palo Alto: Stanford University Press, 1971.

Keep, Josiah, revised by Baily, Joshua L., Jr. *West Coast Shells*. Palo Alto: Stanford University Press.

Kimber, D. C.; Gray, C.; Stackpole, C.; Leavell, L.; and Miller, M. *Anatomy and Physiology*. 15th ed. New York: Macmillan, 1966.

Leet, L. Don, and Judson, Sheldon. *Physical Geology*. Englewood Cliffs: Prentice-Hall, 1963.

Montagu, Ashley. *Human Heredity*. 2d rev. ed. New York: New American Library (Signet).

Pennak, Robert W. *Collegiate Dictionary of Zoology*. New York: Ronald Press, 1964.

Powell, A. W. B. *Shells of New Zealand*. Auckland: Whitcombe and Tombs, 1957.

Rippingale, O. H., and McMichael, D. F. *Queensland and the Great Barrier Reef Shells*. Brisbane, Australia: The Jacaranda Press.

Rogers, Julia E. *The Shell Book*. Rev. ed. Boston: Charles T. Branford Company, 1951.

Romer, Alfred S. *The Vertebrate Body*. 4th ed. Philadelphia: W. B. Saunders, 1970.

Shumway, Waldo, and Adamstone, F. B. *Vertebrate Embryology*. 5th ed. New York: John Wiley & Sons, Inc.

Tinker, Spencer W. *Pacific Sea Shells* [Hawaii area]. Rutland, Vermont: Charles E. Tuttle, 1957.

Torrey, Theodore W. *Morphogenesis of the Vertebrates*. 3rd ed. New York: John Wiley & Sons, 1971.

Trautman, Milton B. *The Fishes of Ohio*. Columbus: Ohio State University Press.

Travers, Louise A. *The Romance of Shells*. New York: M. Barrows & Co., 1962.

Van Nostrand's Standard Catalog of Shells. New York: D. Van Nostrand Co., 1964.

Vilas, C. N., and Vilas, N. R. *Florida Marine Shells*. Rutland, Vermont: Charles E. Tuttle, 1970.

Warmke, G. L., and Abbott, R. T. *Caribbean Seashells*. Narberth, Pennsylvania: Livingston Publishing Co., 1961.

Weier, T. E.; Stocking, C. R.; and Barbour, M. G. *Botany—An Introduction to Plant Biology*. 4th ed. New York: John Wiley & Sons, 1970.

The following periodicals also offered much of interest and value:

National Geographic.

Natural History. Published by the Amercan Museum of Natural History, New York.

Reader's Digest.

Sea Frontiers and Sea Secrets. Published by the International Oceanographic Foundation, Miami, Florida.

An extensive collection of periodicals and bulletins published by different shell clubs in Australia, New Zealand, Hawaii, California, Washington, New York, and Florida.

Index